International Energy Markets

International Energy Markets

Edited by
Paul Tempest

Oelgeschlager, Gunn & Hain, Publishers, Inc.
Graham & Trotman Ltd, Publishers

Published jointly in 1983 by:

Oelgeschlager, Gunn & Hain, Publishers, Inc.
1278 Massachusetts Ave
Harvard Square
Cambridge
MA 02138
USA

Graham & Trotman Ltd
Sterling House
66 Wilton Road
London SW1V 1DE
UK

Available in the USA and Canada from Oelgeschlager, Gunn & Hain, Inc. and in the Rest of the World from Graham & Trotman Ltd.

LCCN 83-60231
ISBN: 0 89946 201 4 (Oelgeschlager, Gunn & Hain)
0 86010 402 8 (Graham & Trotman)

Typeset and printed in Great Britain.

Contents

Preface

'The view that market forces (albeit ill-defined) will once again dictate the future evolution, steady or volatile, of the oil and energy markets seems to me to be not just mistakenly simplistic, but potentially dangerous.'

Sir Peter Baxendell at the 1982 Cambridge Energy Conference

World energy markets are at a new watershed. In some unexpected directions, exciting prospects are opening up; in others, the storm-clouds gather. The balance of power is changing. Inter-fuel substitution and conservation are beginning to feed through at a point when the rationale for them is weakened by a crumbling of the oil price and fragmentation of the oil market. What are the prospects for energy demand, supply and price? To what extent will market forces be inhibited or reinforced by Government action? Will the relaxation of energy constraints lead to a surge in global economic activity and a further swing in the supply/demand balance? To what extent are new market mechanisms being developed? Will they help engender stability? Is the current rationing of energy supply by an artificially inflated marginal price in fact throttling off investment in new and potentially cheap energy which, if abundant, offers the greatest hope of breaking out of current global economic recession?

These questions are at the heart of the current global macro-economic debate. They are inextricably linked with local, national, regional and international politics, and can therefore only be satisfactorily addressed by bringing together a very wide range of experience and expertise.

In Cambridge, UK in June 1982, 240 top-level representatives of government, industry, the leading research institutions and the financial sector from 26 countries were presented with the above questions and then locked into debate of these issues for three days. This volume reviews the main contributions to that debate, the subsequent discussions with lead speakers and session chairmen, and assesses the underlying conclusions. It seeks to provide an authoritative framework for the further analysis of change and potential change in international energy markets and to summarise the direction in which some of the most advanced thinkers in the field are beginning to direct their operations, planning and research. Some marked differences of opinion are apparent, particularly between the international political strategists and the market operators.

The technical work of the Conference, particularly the various energy modelling sessions, those papers and sessions devoted to individual economies and a large number of papers on energy in the developing world, have been omitted from this volume in order to secure an element of balance. The papers and abstracts have been rearranged to provide some thread from a diagnosis of the nature of change in the markets, through the likely response of Governments, international agencies and the private sector, and a regional analysis of the part played by energy in world power politics to a brief review of those risks and opportunities in prospect where solutions may well be found.

One major international policy issue surfaced repeatedly both in the formal debates and less formal exchanges. Ten days before the Conference started, the US Administration extended its embargo, announced in December 1981, on the export of oil and gas technology to the USSR, to include foreign subsidiaries and licensees of American firms. At the time of the Conference, therefore, each of the European countries was in the throes of reassessing its energy policies and of formulating a commercial response. While press attention was focused on the legal issues of American subsidiaries in Europe being caught between two sets of conflicting legislation, the long-term impact of the embargo as a deterrent to overseas investment and world trade was beginning to dominate policy discussion. Much turned on a division of opinion concerning the need for Siberian gas in Europe. Is Western Europe indeed headed for a major energy supply deficiency? What is the prospect of a serious imbalance within the Communist world? What level of energy import dependence is tolerable? What planning parameters should be placed on energy pricing over the 20–40 year life of the gas lines currently under construction? These questions lie primarily within the competence of energy economics and are addressed in some detail in this volume. The embargo extension was lifted on 13 November 1982, but the issues it raised remain central to international political and economic relations.

The issue of the role of Government in international energy markets, discussed succinctly by Sir Peter Baxendell in his opening address, became the central theme of the conference. It was dominated by the current market-orientated stance of President Reagan and Mrs Thatcher, as expressed here by their respective Secretaries for Energy against a background of increasing Government involvement in attempts to salvage or abandon a number of mega-projects in the energy sector, notably in Canada, the United States and Europe. In the area of consumer/producer relations, continuing hostilities between Iran and Iraq and fears that the Israeli entry into the Lebanon might again trigger the use of an Arab oil boycott in one form or another, stimulated lively sessions on contingency planning within OECD Governments and the wider economic effects elsewhere.

Forces and Forecasts

The almost total failure of energy economists to predict the major developments of the seventies has been generally attributed (by them) to the increasing politicisation of oil issues and the unpredictability of Government action and reaction. In his introductory address, Sir Peter Baxendell runs through the main caveats which must predicate any analysis of energy markets such as this, dismissing the simplistic examination of market forces. He doubts whether Government intervention is at all predictable. While the market provides signals, it cannot provide answers. Obscurity, he argues, will therefore always rule in those segments where politics are paramount. 'Nonetheless, elsewhere', he argues 'we may still benefit from serious research, informed analysis of the economics of the system and perception derived from experience'.

Supply Uncertainties

The development of international gas supply and demand is clearly a, if not the, major energy opportunity of the next two decades. Yet, as Professor Adelman concludes, it is essentially a matter for intergovernmental negotiation; alignment with the oil market will be neither easy nor necessarily relevant. The current fragmentation and regionalisation of the international oil market described by Professor Odell may be no more than a temporary phenomenon of global economic recession, but some of the developments appear to be of much more fundamental character and to imply a loss of optimisation through market pricing and a diversion of exploration and development away from some of the most promising low-production cost opportunities.

The high hopes of the seventies for a rapid energy substitution of coal and nuclear energy for oil and gas have faded rapidly. Recovery of momentum following the Three Mile Island accident has been generally slow, despite, as Fritz Lucke concludes, considerable Government effort. World coal trade, similarly, as reviewed here by Rajendra Pachauri and Walter Labys, reveals not merely the frustration of inflated hopes but the inability to sign long-term supply contracts without the complementary commitment of consumer Governments to increased investment to prepare domestic industry to receive and utilise the coal.

The Changing Parameters of Demand

Energy demand in the industrialised world has flattened to a far greater extent than was generally thought possible five years ago. Energy economics has been mainly concerned with separating the effects of global economic recession from the evidence of increasing energy efficiency. This was a major theme which extended over many Conference debates in addition to the plenary and parallel sessions devoted to it. In this volume, we are privileged to publish, almost in full, Sam Schurr's award-winning study of Energy Efficiency and Production Efficiency delivered at the Cambridge Conference. In it he concludes that advanced technology, computerisation and robotics may be just about to provide strong stimulus for radical changes in the production and consumption of energy worldwide. To substantiate this theme we have included abstracts of detailed papers covering the potential for homeostatic energy demand control and pricing (by the Ewbank Group) and a definitive analysis of the choice between Input and Output Energy Demand measurement (by Professors Ernest Berndt and Campbell Watkins). The question of how much we have saved and might save by conservation and increased investment in the domestic sector is covered, in abstract, in statistical terms by Lee Schipper and Andrea Ketoff and in architectural terms by Professor Patrick O'Sullivan and Robin Wensley.

The Future Oil Price

Oil price forecasting continues to play a central role in the formulation of all energy policy and energy investment planning. It is also an essential component in macroeconomic analysis and policy. From the crude single-line extrapolation forecasts which dominated the sixties and the proliferation of scenario planning in the seventies, energy forecasting of demand is developing, with the aid of much more sophisticated computerisation, into a system where the macroeconomic feedback of a range of assumptions begins to give much more plausible planning parameters. In this volume, Henry Jacoby and James Paddock of the Massachusetts Institute of Technology present their concept of an 'oil-price window'. Computer technology is indeed on the brink of being able to convert any set of assumptions into such an easily understood 'oil-price window' at any point in future time so that different forecasts can be readily and easily compared by both user and modeller. The same three-dimensional techniques are being similarly applied in resource depletion economics, so that long-term oil price forecasting will be subject increasingly to much more rigorous investigation in terms of resource depletion. In the shorter-term there are even more exciting developments, where sophisticated oil-price forecasting is becoming an operational tool in the new futures markets for oil products and where the commercial expression of any future energy price can be fed back into the forecasting model. In summary, we are emerging from the world of the modeller, submerged in computer print-out, guessing blindly and struggling to communicate his own narrow perception, in his own terms, into a world where the results of all the major forecasters become widely and instantly available and where they can be consulted easily in a comprehensible and comparable form. All these will be revised and updated on a continuing basis, so as to provide specific answers of ranges of probability for specific policy and operational decisions.

Oil Refining and Substitution Economics

Most long-term energy supply contracts take account in escalation clauses, of the future price of crude oil and oil products. Yet, as John Culhane explains here, the structure of oil refining is changing so fast with a changing input mix and upgrading of refining and hydrocracking capacity on the supply side and new competition from alternative fuels and changing markets on the demand side, that historical price relationships between the main oil products are of little use. Whether, for example, the production of heavy fuel oil, with its declining share of the barrel, will outrun market demand and become a price-maker or a price-taker *vis-à-vis* its alternatives defies prediction. The genuine difficulty of selection of the appropriate indexation in major energy supply contracts lies behind much of the current delay and confusion in this field which is ascribed all too readily and wrongly to commercial intransigence or political machination on the part of the producer or consumer.

The development therefore of contractual terms which are flexible and yet express the mutuality of producer and consumer interest in a form acceptable to the world capital markets is likely to become a most pressing priority for Governments and industry throughout the world.

The pricing of oil, however, shows every sign of becoming increasingly complex. In Part XIV Fereidun Fesheraki and David Isaak outline the likely impact of new Gulf oil and petrochemical products on world markets. Robert Deam and Michael Laughton take the theme much further seeing a solution between the OPEC producers and OECD consumers in the form of downstream investment by OPEC through processing deals and equity participation.

We also include in abstract an important paper by Bruce Netschert which argues that the oil industry, caught with excess refining capacity, often has little interest in developing or even encouraging gas development or those processes, such as methanol, which by-pass the refining process, and which in many countries, may nonetheless be an important component in the energy mix.

Comparative Licensing, Leasing and Taxation

The 1982 Cambridge Conference drew on the conclusions of the previous conferences in Washington (1979), Cambridge (1980) and Toronto (1981). They demonstrated vividly the value of commissioning comparative studies of national policies on energy exploration, development and production. Anomalies in national policy on licensing, taxation and participation could be highlighted by comparison with other regimes and the penalties of extreme positions quantified. Three important papers were therefore commissioned for the Cambridge Conference. Walter Mead, Asbjorn Moseidjord and Philip Sorensen conclude here that Governments must choose a market-oriented leasing system to achieve maximum economic efficiency and an effective Government share of the economic rent.

In the field of taxation, the issue is similarly whether certain fiscal regimes, particularly those in Canada and the North Sea which are highly tuned to capturing a high economic rent, are in fact causing long-term damage by deterring marginal investment. We therefore include the summaries of the very detailed paper by Alexander Kemp and David Rose which compares the effect of taxation on petroleum exploitation in 15 different fiscal regimes, and of a paper by Timothy Greening of Chicago which examines the non-tax barriers to exploration, and which produces a useful hierarchy of constraints on such investment.

Straight comparative studies of one facet of Government policy are not enough. Any valid comparison has to review all aspects of Government involvement and place it in its broad economic framework.

The Role of Government

In the second major part of this volume we reproduce the major policy statements of the United States Secretary for Energy, the United Kingdom Secretary-of-State for Energy, the Executive Director of the International Energy Agency and the Director-General of the Energy Directorate of the European Commission. We are very grateful to Mr Edwards, Dr Lantzke and M Carpentier for travelling to the Conference despite very heavy work schedules and for debating the main issues both in formal session and over luncheon and dinner. Mr Lawson addressed the Conference on 28 June 1982, the day on which the Oil and Gas Enterprise Bill received the royal assent and became law. His explanation of this key legislation, which provides for the privatisation of most of the British National Oil Corporation and the reintroduction of commercial competition into the United Kingom gas industry is widely regarded as a turning point in the balance between public and private sector activity in the United Kingdom. During his address, Mr Lawson also announced the support of the United Kingdom Government for a land-bridge to deliver Norwegian gas through the United Kingdom and a Channel or Southern Basin gas line to the markets of continental Europe. The possible integration of UK gas and electricity supply with the networks of continental Europe also therefore became a theme of debate. Opinions differed widely on whether this was in the short-, medium- or long-term interest of the United Kingdom, Norway and the countries of continental Europe.

The Response of the Business Sector

The effect of increased Government intervention has led to marked changes in the structure of energy trade. Government-to-Government deals using state corporations or their agents now cover a significant share of oil trade. The ability of the multinationals to finance, with a weakening cash-flow, the bulk of exploration and development is, as argued by John Mitchell, therefore thrown into question, as is their ability to compete in many areas with Government bodies whose guarantees, privileged marketing position or wrapping-up of energy contracts in broader trading, defence or development packages, obscure the agreed price. Equally, however, public corporations competing in an international market have to behave to a large extent as private sector entities. If excessive direction is given to their activities such as severe Government limits on their activities overseas or on their access to external markets, they are placed at a distinct competitive disadvantage. Whether in the public or private sectors, the most important corporate resource is always the skill, flexibility and initiative of the staff. Its successful motivation is the prerequisite for corporate efficiency and effectiveness, in both the public and private sectors, which must remain at the forefront of any national energy policy.

Contingency Planning

The question of energy supply discontinuity and the need for contingency planning and preventative policies in the OECD countries has stimulated a wide variety of proposed solutions. We include here Philip Verleger's tour d'horizon, which together with the conclusions of John O'Leary point towards an oil tariff. In reply, the argument of Lawrence Jacobson and Ralph Tryon of the Federal Reserve Board concerning the external and internal effects of oil import tariffs, emphasises that exchange rate effects tend to dilute much of the effect and create serious repercussions for the domestic economy.

Policies in the disposal of commercial and Government stocks in times of supply shortage occupied several sessions, including one devoted to the development of oil futures markets. So far, however, these markets in oil products in London and New York have been used mainly for small-scale hedging.

Communist Bloc Energy

As explained above, Conference debate centred on the Siberian-European gas deal and the degree of future European energy and economic dependence. The European perspective is set out by M Carpentier in Part IV and, in abstract, in Part XIII by Ferdinand Banks of the University of Uppsala.

Tony Scanlan's paper, reproduced here, leads to one major conclusion. Eastern Europe is shortly likely to be increasingly energy deficient; to the extent that it can turn partly to the USSR, it is likely to have to pay world prices. The consequent compulsion to seek Western sources of energy, notably OPEC, trading East European technology and skills and bartering industrial goods and services to secure energy price discounts, could become a strongly destabilising force in international energy markets over the next two decades.

Towards Self-sufficiency in North America

The progress made by successive United States Administrations to lessen oil import dependence is charted here by John O'Leary with a plea for a more positive price-active energy policy. His prescription of the need for a flexible oil tariff should be read against the conclusions of Lawrence Jacobson and Ralph Tryon in Part IX. The abstract of an elegant analysis by John Fraser of the gas price mechanism in the United States is also appended. It points out that Alaskan gas and LNG imports might be backed out, on market price reasons, by Canadian and Mexican supply.

Pricing Problems in the Pacific Basin

The difficulties of Australia, rich in energy resources and within reach of Japan and other profitable markets illustrate, in the papers included here, many of the uncertainties which hamper development in the area at present.

The West European Supply Dilemma

The main argument for Western Europe is set out in M Carpentier's analysis in Part IV. Here Ferdinand Banks reinforces the argument and Melvin Conant poses a wider question—whether, in the long run, there is any hope of bringing potential OPEC gas supply and European energy demand into a much closer political and commercial relationship?

New Strategies in OPEC Planning

The injection over the next five years of massive volumes of oil product and petrochemicals from new refineries and industrial complexes in the Gulf poses serious questions for the international markets. The close involvement of the oil multinationals in this new source reflects their long-term interest in securing access to crude and it appears unlikely that the marketing of Gulf crude can for long remain unlinked with the marketing of Gulf products and petrochemicals. Here Fereidun Fesheraki and David Isaak set out the most likely numbers and consequences, while Robert Deam and Michael Laughton explore the possibilities of much greater OPEC involvement in OECD refining through processing deals and 'share farming'.

Breaking International Deadlock in the Developing World

Strangled by debt servicing, much of the developing world is beginning to pay the price of relatively fast growth in the seventies. We reproduce here a symposium of ideas to break out of the deadlock which is inhibiting energy development in these areas.

Energy and Economic Growth

Without global economic growth, energy development will be blighted; most trace the seeds of the current global economic recession to the oil price discontinuities of the seventies. Here we include abstracts from the major paper by Nobel prize-winner Wassily Leontief and Ira Sohn and a selection of other key contributions.

Conclusions

The overall conclusion stands in marked contrast to the guarded optimism of the last Cambridge Energy Conference held in June 1980. Since then, much of the expectation of a rapid development in non-oil energy, particularly coal and nuclear, has evaporated; demand forecasts have been revised progressively downwards; problems of surplus refining capacity, a declining share of oil in primary energy and a persistent slackness in oil prices, together with continuing Government concern for security of supply, all indicate a slackening of global energy trade and a regionalisation of energy demand which may have far-reaching implications for world politics and further global economic growth. If we accept the drift of this diagnosis and argument, we must begin to come to terms with a number of major stress-points. In the forefront are the imminent exposure of Eastern and later Western Europe in terms of energy supply; the potential isolation of the Gulf oil producers; and the lack of any coherent strategy for the developing world.

In the longer-term, the current slowing down of investment in energy may inflict even heavier damage on the global economy. If, indeed, individual Governments again react to supply problems with measures which distort the international markets and vastly and unnecessarily increase the resource cost of energy, the prospects for political co-operation and global economic prosperity are bound to be placed in jeopardy. These problems do not appear to be insoluble.

A starting point would be much greater appreciation by Governments of the changing nature of international energy markets and an awareness by all of the need to cope more flexibly in new energy investment, finance and supply agreements with market fluctuations and uncertainties.

SECTION 1

TOWARDS A DIAGNOSIS

PART I

International Energy Markets

Chapter 1

Forces and Forecasts

*Sir Peter Baxendell**

There seems to be a growing tendency among commentators and policy makers to believe that many of the forces which made the oil markets turbulent in the seventies—OPEC policy, consumer country vulnerability and the imposition of political interests and influences—no longer apply, and that 'market forces' (albeit ill-defined) will once again dictate the future evolution, steady or volatile, of the oil and energy market. This seems to me to be not just mistakenly simplistic, but potentially dangerous.

An edited action re-play of the international oil business during the past 10 years would highlight two major and largely unforeseen shocks (in 1973/4 and 1979/80), some violent short-term reactions, and a progressive but profound structural change. I want to discuss with you the forces at work then and now. My main purpose is to point up the dangers of misreading causes and effects in times of uncertainty—especially when conclusions are drawn from incomplete analysis and are then injected into forecasting.

*Chairman, Shell Transport and Trading Company.

COMMON FEATURES OF THE 'CRISES'

The two so-called oil crises of the seventies had certain features in common. In both cases, market conditions were such that the balance of supply and demand for oil in international trade was already precarious: there was rising demand and either a growing inability or a diminishing inclination to supply that oil. Then supply was severely disrupted because of political events—firstly the Arab-Israeli war, and six years later the revolution in Iran. Both emergencies were followed by a period in which the duration of the emergency was uncertain, and the intensity and significance of its impact on oil supply was a matter of speculation, giving rise to continuing anxiety among importers. In both cases there was a shift of influence over the supply and price of oil towards those who control supply—something which permitted them to achieve policies on supply and price which reflected, as we know, a whole range of individual national and international political and economic objectives.

PRODUCER ACTION

Let me remind you of just a few of these influences. As prices rose rapidly, the priority of maximising sales and revenues gave way, in many cases, to the concept of extending the economic life of finite reserves. Government-to-government deals took precedence over the traditional commercial channels of oil supply. Politically motivated embargoes on the destination of oil were introduced or extended. Premia were imposed selectively on individual transactions and there was unilateral amendment or cancellation of oil supply contracts. There was no longer any coherence in crude oil pricing, with differentials out of line with their value in the market place, and with spot market prices soaring in response to actual or perceived shortages, so that the whole price structure was dragged upwards through the influence of a very small proportion of the trade.

CONSUMER RESPONSE

As the balance of power shifted, consumer governments realised the vulnerability engendered by their dependence on oil imports, and felt it increasinly necessary to become politically and bureaucratically involved in the supply and pricing of oil. Again, there are many examples of this involvement, and I will quote just a few. Price controls were extended or introduced, especially in the 1973 crisis, and were often applied in ways which seriously inhibited the market from performing its proper role. Stock administration became subject to increasing government interest, so statutory levels were influenced by national security considerations rather than operating requirements or commercial criteria in financing them. There was a notable extension of national oil companies' involvement in the international oil markets and many continued to operate outside the discipline of normal commercial criteria. And there were bilateral diplomatic negotiations to secure privileged supplies in response to special aid or trade treatment. These actions and reactions were sanctified, so to speak, in the declarations of the Venice Summit, following which nearly half the communiqué was devoted to energy matters and intentions.

I have no quarrel with those who say that the 1970s were a period in which the international market for oil was especially heavily influenced by non-market forces. But I begin to part from them when they carry on to say that this phenomenon was nothing but a temporary aberration, and that now the 'market' has taken over again to drive us into the future.

Perhaps I should try at this point to clarify what I mean by 'market forces'. Grossly oversimplifying, I am talking about the free interaction of cost and price upon supply availability and demand, in a way which determines the structure of the market and its subsequent evolution through the investment behaviour of the participants.

MESSAGES FROM THE MARKETS

Let me play the devil's advocate—from my point of view—for a moment, and list some of the reasons put forward for thinking that the traditional concept of the market may once again be in charge.

Firstly, there is the current drop in demand for oil—something to which I will return later. I think it is widely accepted that the bullish projections of future oil demand that were common in the early seventies are no longer valid. They have been replaced by much more modest expectations which, in turn, tend to remove people's worries about *technical* availability of supply and also about the life-span of known reserves.

Secondly, there is the present surplus of OPEC technical export capacity relative to expected demand for OPEC supplies over the foreseeable future.

Thirdly, there is the development of non-OPEC supplies. As you know, non-OPEC crude oil and liquids are increasing their share of the non-communist world's requirements: they accounted for 48% in 1981 as compared to 40% in 1979, and the OPEC share is expected to drop further this year.

These are the factors which reinforce the conviction of those who believe that 'cartels can't last', and who argue that the intervention of forces other than traditional market ones had their sole origin in OPEC's actions when OPEC enjoyed most of the power. They assume that that 'power' has now shifted back to the consumer and that as demand in the consuming countries continues to fall, the pressures which disrupted market forces during the 1970s are progressively lessening. As a result, we now hear almost as much discussion on the possibility of a collapse of oil prices as only a year ago was devoted simply to speculation on how fast they would rise.

I fear, however, that this could well be another example of the sort of myopia that so often afflicts those dedicated to gazing into the future—the tendency to adopt present perceptions and feelings as the soundest basis for a view of the future, a view which thereafter is presented with great conviction. We, however, do not believe that events of the 1970s resulted from a temporary aberration, or that the energy problem has gone away. Let me explain our doubts and tell you why we have no more confidence in forecasting the future now than the low level of confidence in that art we reached through the traumas of the last decade.

POWERFUL NON-MARKET FORCES

First and foremost, the supply of oil to the world market is still highly vulnerable to political action or accident. In the period following the first oil crisis, there were two dominant OPEC actors—Saudi Arabia and Iran. Both were concerned with maximising the revenue from their primary resource but also saw that their own interests lay in preserving reasonable stability in the world's energy supply. The post-crisis adjustment of supply to fluctuating demand was considerably facilitated by this, although it is significant that crude oil prices showed little downward flexibility when balance was restored. There is now only one dominant producer in the unstable Middle East scene, with Saudi Arabia producing around 40% of OPEC oil compared with only 25% in 1973/4. To put it another way, the international oil industry still relies on one source for up to 30% of its supply, so developments in, and the policies of, that country will continue to be a major influence.

It must also be obvious in today's economic and political environment that, at its present price, oil remains of major significance both to importers and exporters. Energy costs and security have and will continue to have a major impact on national economic policies, for good or ill. They impinge on almost every aspect of national life, from inflation and balance of payments considerations through industrial costs and structure down to the lifestyles of individuals. Many legacies of the oil shock still remain. Governments of energy importing countries (and some with domestic surplus) have become increasingly interested in developing indigenous supplies of energy, even in comparatively small quantities, in order to improve supply security, employment and the balance of payments, and many such ventures require subsidies. On the international scene, the IEA remains in place. And on the producer side, it is worth remembering the concerted OPEC reaction concerned with the impact on Nigeria when the market brought about a severe reduction in its offtakes. Add to that the unexpected degree of agreement on supply and pricing achieved by OPEC members in the spring and reasserted at the Quito meeting—it is too early yet to assess whether those decisions will result in total success, but no one can deny that they did indeed have an effect on market behaviour and perceptions.

Let me now go back to examine some of the judgements I mentioned earlier as being invoked by those who see the market as having taken over the driving seat for the energy industry.

NON-OPEC SUPPLY INELASTIC

First, we could well be misled by the fact that the proportion of non-OPEC oil in international trade has recently been rising so significantly. This phenomenon results very largely from the fact that OPEC sources bore the brunt of the fall in demand, while the major new non-OPEC export sources, principally the UK and Mexico, have continued to produce at levels relatively close to their technical potential. Apart from other considerations, the latter have not been so constrained by attempts to maintain prices. When demand recovers, with economic upturn, the increment will largely have to come from OPEC sources. Increasing technical potential from non-OPEC sources—even when major reserves have been established, as in Mexico—will take considerable time to develop, in view of the sheer complexity of offshore development. Furthermore, the marginal technical cost of traditional OPEC supplies will be much lower than that of incremental supplies from the new provinces such as the Far North and the North Sea. Thus a fundamental longer-term shift in the OPEC/non-OPEC balance will take many years to achieve, and, if achieved, is likely to be relatively short-lived given the immense OPEC reserves.

Similar considerations apply to expectations for the contribution from the US as a non-OPEC source of oil. Undoubtedly the deregulation and 'internationalisation' of oil prices there provided a major market incentive for exploration and development. However, except for the expensive Far North, the United States is one of the most mature of oil provinces; it is probably not exaggerating too much to say that the market response there has primarily been in arresting the decline in output and reserves replacement over recent years. Important though this is, it cannot be said that the restoration of market prices in the US will do very much in the short term to influence the supply side contribution to international trade from the United States. (It is interesting, nevertheless, to observe the major cut-backs in US exploration efforts this year in response to recent lower expectations for crude prices.)

In short, I am urging that we should not over-estimate the price elasticity of supply available from non-OPEC sources in the short term.

THE FUTURE DEMAND FOR OIL

Turning now to the assumptions about international oil demand, one major element in our analysis is the influence of substitution by alternatives to oil—in the short term, coal used as coal, and natural gas; longer term, less conventional substitutes such as coal gasification and liquefaction, and tar-sands and shale oil. We all know that the nuclear contribution in most countries is being inhibited by long lead times and political and environmental constraints. The short-term substitution of either coal or natural gas for fuel oil depends not only on comparative cost calculations, which are seldom straightforward, but also on infrastructure limitations, security and diversity of supply considerations, as well as the consumer's willingness to make major capital investments. To state the obvious, it takes time, money and commitment to make the transition from an energy system heavily dependent on oil (mainly from OPEC) to a broader, more diverse supply base. To illustrate, over several years of policy support for coal, its contribution has grown at a steady 3% per annum—no dramatic shift of emphasis.

As for the longer-term development of alternative energy supplies, major new projects are unlikely to be undertaken unless there is a general perception of the need for them, and an apparent economic incentive for the investor. Unfortunately, the present climate is hardly conducive in either sense, for the necessary decisions are inevitably affected by the very perceptions which I am trying to moderate today, while the lack of consistent, encouraging energy policies in many consumer countries is a major problem for those who would otherwise be willing to take the long-term risk of investing in the more costly alternatives to conventional oil and gas projects.

Looking further ahead there are few, if any, non-conventional alternatives to oil which at present are estimated to cost less than $40 a barrel in 1982 dollars. The present official price of conventional oil is still considered to be under pressure despite apparent agreement among OPEC members. On the demand side, the costs of conserving oil by investing in more efficient equipment vary fairly widely, but on broad general assumptions for both domestic and industrial efficiency levels they tend to cluster around $15 to $20 a barrel of oil saved. Thus the stimulus to initiate large scale development programmes of synthetic alternatives to oil—projects which have long lead times and demand considerable research and development expenditure with concomitant risk, is being eroded.

The final factor which leads me to doubt the premise of an energy evolution brought about largely by the influence of traditional market forces is the still uncertain nature of demand projections for oil in uses which are not readily subject to possible substitution. To start with, there are a wide variety of assumptions on when we can expect a widespread pick-up in industrial and economic growth, and until that occurs it is almost impossible to quantify the degree to which oil demand has been affected by the recession and continuing stagnation in world economic activity. When economies begin to revive we will also be able to judge with rather more accuracy how much the fall in oil demand has resulted from a genuine and permanent change in consumption patterns and how much from temporary 'belt tightening'.

Clearly while recession persists in the main oil markets and market adjustments to the present level of energy prices are still flowing through, the consumer has a greater say in the market than has been the case for some time, and—again at present—other players are dancing more to his tune than he to theirs.

But we should remember that there has been no actual shortage of crude in the market for some years and, perhaps even more important, no perception of actual or prospective shortage since 1980. And yet early last year the marker price went up in dollar terms, and even more in terms of European currencies as the dollar hardened. And it went up again last October. Furthermore, since we now appear to have passed through the period when the short-term survival of the $34 marker price was in serious doubt, the only remaining significant feature reflecting the shift of influence to the consumer in the international market has been a reunification of prices when those unreasonably out of line came back into line with the marker. That in itself is inadequate evidence that the 'normal' influence of a free market is the predominant force.

The marker price provides an interesting illustration of the limitations on the power of market forces in the international oil scene. Whether the marker price is $34, $30 or $25, the marginal supply generates massive rents to the key suppliers—rents which are now largely taxed away if the initial beneficiary is private industry; nevertheless OPEC producers are now selling less than half of their available production rather than reduce their price. This is in sharp contrast with the situation in downstream markets where—taking marginal refining margins as an appropriate equivalent—a substantial surplus in capacity over demand causes such competition and price pressures that much of the downstream industry is struggling and some are falling by the way. *That* is the traditional market at work.

To sum up, the complexities of all the interacting forces make me frankly concerned about over-simplified interpretations of the structure and prospects for this international industry. I think the search for a simple model of the structure of the oil industry is wholly understandable, if only because so many people concerned with it—academics, financial analysts and policy makers both in governments and in the industry itself—would like to have a secure base for predicting the future. But I am convinced that they will continue to be frustrated in their search because the real model is far too complex to be captured by equations. So often the crucial factors are relationships deriving from the unquantifiable logic of politics and power, and are based more on inadequate perception than on reality.

If we carefully differentiate between those segments of our analytical interest where obscurity will always rule and, on the other hand, those where we may still benefit from serious research, informed analysis of the economics of the system, and perception derived from experience, we may find that we are less vulnerable to future shocks, and that the crystal ball we would all like to have will be somewhat less of a soap bubble.

PART II

Supply Uncertainties

Abstracts (see Section 5)

Chapter 2

The Changing Structure of the Oil and Gas Market

M. A. Adelman*

In my judgment, there is no world market in gas, and in few if any countries is there any domestic market. We do have many fragments, i.e. individual bargains whereby pipeline gas is bought and sold at some price. The product is uniquely homogenous: methane, at a given pressure and temperature. But gas does not fetch anything like the same price throughout any area, once we make allowance for transport costs.

One reason for the lack of a gas market is that there are almost no spot sales, nor any arbitrage, either by producers, consumers, transporters, or others. This is very much unlike the oil market, where a dense network of buyers and sellers has long made it impossible, for example, to have a boycott of one or few buyers. The oil boycott of Italy in 1935 failed, as did the 1973/74 boycott of the United States and Holland, because any price rise, or expected price rise, in one or two countries, promptly re-directs cargoes: and a small diversion or swap is enough to even out prices. (Of course, the 1973/74 production cutback quadrupled the oil price *level*, everywhere.)

Natural gas transport costs more than oil transport: roughly three times as much by land, and six times by sea. The volume of world

*Professor, Department of Economics and Energy Laboratory, Massachusetts Institute of Technology.

trade is still very small, and such a re-direction of supply would be a major effort and de-stabilizer, instead of a stabilizing fringe to the market. The world industry is small and immature. But the United States accounts for half of total marketed production, and has a dense pipeline network, so that a surplus or shortage is quickly transmitted over a very wide area. Yet prices vary enormously and irrationally, by a factor of about four.

The reason, as everyone knows, is highly detailed field price regulation of natural gas, which has been effective for roughly twenty years. The Natural Gas Policy Act of 1978 has given regulation a fearsome complexity; like the peace of God, it passes all understanding. Prices are regulated in about 22 categories, and even over this number there is disagreement. All we can say with any assurance is that most allowed prices are well below the market clearing level. The excess demand is all focussed on to the unregulated sectors: very deep gas, and some newly discovered gas. Buyers are willing to pay more for unregulated gas than they can charge upon resale, because they 'roll in' the high-priced gas with low-priced gas, and charge their customers the average of all gas. Hence the proposals for gas from Alaska whose cost would probably far surpass $10 per mcf; it is to be subsidized to make it affordable by depriving producers of the incentive to produce cheaper gas.

This rolling-in has led to serious disagreements with Canada and Mexico, who resent the fact that they are not permitted to collect the ultra-high prices; their gas is as good. Very little Mexican gas is therefore sold in the United States. But much Canadian gas is largely unsold even at the supposed unduly low price. Canadian gas (it had seemed) was too scarce and precious to ship south of the border. Only in May 1982 did the National Energy Board of Canada finally authorize more shipments when it was painfully obvious that the US market would not take even the smaller amount they had previously authorized. Canada is even looking seriously at exports to Asia and Europe, which would fetch them much less. But in the United States, pipelines which were formerly willing to pay $9 to $10 per mcf for rolling-in, are now taking advantage of their contract options, refusing to pay more than $5 or $6 because they can no longer obtain enough from consumers to support the higher prices.

The gas 'surplus' is today worldwide. Development is being reduced, cancelled, or postponed. Gas reserves continue to grow, and few would doubt that the development process, and exploration for gas as such, would make reserves grow faster, since most deposits have thus far been found incidentally to the search for oil. Everywhere there is the same reluctance to see that the product is simply overpriced, given current prices of oil, coal, and nuclear power. Gas is no 'premium fuel' compared with oil. On the contrary, it cannot be used to make gasoline, the most 'premium' of products. Its one advantage is that it costs less to use for steam raising than oil or coal because boilers are simpler and there is no need for anti-pollution measures.

UNIFIED GAS PRICING: A CHIMERA

Were there an effective competitive market in gas the price would be determined by the intersection of supply and demand. With very little gas, it could be used only as petrochemical feedstock, at a price comparable to naphtha. With more available, its marginal use might be in competition with light fuel oil for home heating. With much more available, as is now the case, its marginal use is with low-sulfur residual fuel oil and with coal, naturally low-sulfur or de-sulfurized. But the supply of residual fuel oil is bound to shrink because its sale will be unprofitable, given crude oil prices. As refineries close down, and remaining capacity emphasizes ever more severe cracking, residual will be sold for what it can bring, which will increasingly approach coal prices.

But, gas does not stand in any rational relationship to the prices of competing fuels. Hence there is wide currency for a bizarre notion: that gas should sell at approximate thermal parity with crude oil, at the point of production. When a market existed in the Southwest United States, decades ago, gas did fetch an approximately single price, adjusted for transport. That price was a small fraction of the price of crude oil.

The long run marginal use of natural gas in North America and Europe is somewhere in the range between $2 and $4.50 per mcf (thousand cubic feet), but closer to the lower end. That would be a competitive solution. A rational monopolist of natural gas would discriminate in price, wherever he thought he could control the re-sale, and charge each class of customers or each consuming area, what that class or area could afford. Prices in Europe are far above competitive but neither are they monopolistic.

The proposed sale of Soviet gas will be apparently around $4.50, at the high end of the range. But this estimate is shrouded in considerable mystery on details which will almost certainly change the total appreciably. By comparison, the subsidized credit on equipment sales to the Soviet Union seems trifling. The total cost of the project is estimated at $10 billion to $15 billion. The amount loaned can hardly exceed $7 billion, for eight years. The subsidy amounts to an additional 20 cents per mcf, probably less.* This is a small part of the margin for error of the gas price for the first eight years. Meanwhile, there is serious trouble in arranging for disposal of the Soviet gas; so far, only about half has been spoken for.

Algerian gas landed in Europe will apparently be around $6, about a third higher than Soviet gas. The French are overpaying for Algerian gas just as in the 1960s they overpaid for Algerian oil. I suppose they will accomplish as much now as they did then. The Algerians are demanding the same price from the Italians, who complain that if they had known the Algerians would so increase the price they would never have built the pipeline under the Mediterranean, a great engineering achievement which is now a bargaining disadvantage. The Algerians have a contract with Panhandle Eastern in the United States; the price was adjusted upward dramatically, but they have demanded more, and have discovered 'technical difficulties' to halt deliveries until they get the price as much as the ultra-high prices mentioned earlier—which nobody is getting any more! Consumer demand has been a closer limit on Algerian demands than have contracts, those 'inky blots and rotten parchment bonds', as an Englishman described them.

Even a competitive price level would allow handsome to huge profits producing and transporting additional gas. Here governments are at an advantage compared with private companies. Since they pay no taxes, they can look straight through the fiscal regime, compare prices with real economic costs, and perceive that prices are usually a multiple of costs. Various reasons hold them back and I cannot tell how long they will be effective. Some like Norway fear the impact of additional revenues upon the macroeconomy. The remedy is very simple but apparently impossible—use the capital market to park excess funds until the best time for their use. The legend of 'depreciating paper' is durable and unfounded: anyone can preserve capital by buying the short term obligations of a stable government, and one who invested in US Treasury bills throughout the 1970s would show

*Assume $7 loaned, and repaid after 8 years, at a concessionary rate of 7.8 percent instead of the non-concessionary OECD rate of 12.4 percent. The repayment will be $\$7(1.078)^8 = \$12.8$. But the present value, properly discounted, is: $\$12.8(1.124)^{-8} = \$5.0$. Hence the subsidy is the difference between $7 and $5, or $2. The gas contract is for 1.4 billion mcf per year for 20 or more years. Then to find the price which would be equal in present value to the subsidy:

$$\$2 \text{ billion} = x[1.4 \text{ billion mcf } (\sum_{1}^{20} (1.124)^{-20})]$$
$$\$1.43/\text{mcf} = 7.3\,x$$
$$x = \$0.196.$$

a modest real rate of return (i.e. net of inflation) on average, though some years would be less, some more. But the most basic obstacle has been the deeply held belief that gas prices must move in step with oil prices, and that oil prices must march 'higher still and higher', evermore. It is time, then, that we looked at the oil market.

OIL MARKET INSTABILITY

To understand the market structure today, we must go back to 1978, when prices were soft and exports disappointingly low. Exports not production are the relevant figure. Internal OPEC sales are not in the world market and have no effect on world supply, demand, and price.

In 1978, prices were soft, because small amounts of competition on price differentials kept seeping into the market. The relative *values* of crude oils constantly change because of constant changes in relative demands for the various petroleum products. But crude oil *prices* are rigidly set out in advance. As companies slowly changed from lifters of equity crude to buyers of crude, they would to some minor extent shift to the better buys. They became less reliable off-takers. Thus fixed prices meant fluctuating market shares for governments. Some were seriously embarrassed by unexpected market share losses. Nigeria in 1966-67 thought it had an arrangement with Algeria and Libya on premiums for light sweet crudes. By the Spring of 1978 they realized that their fellow Africans were undercutting them by shading prices and were in serious financial straits, having overspent their declining incomes.

Millions of dollars and many hours of computer time were spent in the vain attempt to arrive at a 'right' set of differentials. The failure was generally acknowledged by 1978, and the movement began toward direct production control to keep market shares, if not completely stable, at least predictable.

OIL PRICE VOLATILITY

As for the price level it was accepted by all parties in 1978 that the price was going to be raised in early 1979, and indeed spot prices rose late in the year because of this anticipation. Then came the Iranian revolution. It did not produce any crisis. In the first quarter of 1979, oil consumption exceeded production by hardly more than the usual seasonal rundown of stocks, within the range of variation which normal inventories cover. After March, production always exceeded consumption. Also, there were several million daily barrels of unused productive capacity. But these were not to be used. On 20 January 1979—a day to remember—Saudi Arabia cut production from 10.4 mbd to 8 mbd. The cut was only partly restored, to 9.5 mbd in February, and by mid-February the spot price was over $31.

'It's one thing for the Saudis to use the 8.5 million [barrels daily] allowable in normal times to create a nice tight supply situation,' said an oilman, 'it's another to use it to create a world crisis'. Aiming at the 'nice tight' market, they achieved the crisis.

The production cuts had a double impact. Supply was less, but more important was the complete uncertainty of how much would be available month to month, or even week to week. There was therefore an abrupt surge in demand for hoarding over and above consumption. This drove up spot prices and soon official government prices. In the general scramble for short run gains, the governments cut the multinational companies out of third party sales, and permitted them to buy only for their own use.

In June 1979, the Saudis reassured the world that they 'would never allow prices to rise to $20 a barrel'. But in January 1981, when the official price was $32, Sheik Yamani called the price explosion 'another corrective move' like 1973–74, i.e. deliberate and intentional, as it truly was.

Today each OPEC government sets out in advance the amount it will produce in the next month. But any fixed production limit tends drastically to de-stabilize the market, as in 1979. Nobody can fine tune a market with coarse instruments. Insecurity of supply is permanent and incurable so long as the cartel nations fix output in advance.

Oil production figures are a month or more late, and their accuracy is diminishing. Inventory and consumption data even for the OECD nations are six months or more out of date. Even when they arrive, the numbers are inaccurate because nobody knows how much distributors' and consumers' stocks have changed. And outside the OECD nations, there is almost complete statistical darkness. Accordingly the decisions of the Saudis and their colleagues are made in ignorance. Except by improbable chance they will produce too much or too little. In 1979 a group of EEC commissioners toured the Persian Gulf and summarized the aims of the producing nations: to maintain a chronic small deficit. This brinkmanship is sure to give us occasional fear of a shortage, which suffices to produce one.

OIL AND GAS MARKETS ADRIFT

The situation in mid-1982 is like 1978, but much more so. Excluding the Communist countries and OPEC, world oil consumption in 1982 will about equal that in 1971. Talk of 'recession' is misleading; there have been recessions and revivals since 1971. Non-Communist non-OPEC world economic activity this year is up about 35 percent since 1971, hence oil consumption per unit of income is down by a fourth. This downward trend will continue because adaptation takes time. The world has not yet fully adjusted to the 1973-74 price rise, let alone the bigger 1979-80 rise. I think the 1981 decline was exaggerated because of de-stocking by distributors and consumers; the forthcoming recovery will also be overstated by re-stocking.

The increase in worldwide gas consumption since 1972 was very small. It will be somewhat greater in the coming decade, thus cutting farther into oil consumption. The suppliers of natural gas are not the same as the group of dominant sellers of oil. They will drift profitably into taking more of the market from oil. They would sail, faster, if they saw their opportunity as a reduced price and a reliable commitment to supply, but understanding comes slowly.

Moreover, as oil consumption stagnates or declines, non-OPEC production slowly creeps up, and forces OPEC exports down. In April and May of this year they were about 14.5 million barrels daily (mbd), half of the 1979 figure, which was itself below the 1973 peak. I expect OPEC exports will soon revive, but these are waves on a long decline. (Recall that the Central Intelligence Agency in 1977 reckoned 44 mbd OPEC exports 'needed' in 1985.)

OPEC STRATEGY

In February 1980, the OPEC long term strategy committee saw the optimal monopoly price to be just below the cost of synthetics. Hence there was supposedly plenty of unexerted price-raising power. They should now recognize the error: the consumer response is the constraint, and sets the price ceiling. I do not know the location of the ceiling any better than they do. Very possibly they have already surpassed it.

The cartel has wavered between two models. (1) Saudi Arabia lets everyone else produce freely, while it makes up the difference, producing just enough to engender the price best for itself. But (2) it is far more profitable for the cartel nations to agree somehow, however temporarily and irregularly, on a price and profit which is best for the whole group. This is difficult, for no two partners have quite the same optimal price. A far more important problem: in every coalition of one dominant partner and several small ones, each of the small can refuse to carry its share of the burden, knowing that the others can, easily. But the big partner is a prisoner of his size; what he does will make or break the group, and he must rather bear an undue part of the burden than retaliate. So strength is also weakness.

Saudi Arabia has played a strong hand with skill. In 1979-82, as in 1976-77, they 'led the regiment from behind', preferring to under-price and thereby gain increasing market share. As the glut worsened, one no longer heard that they would love to produce much less. Instead, in March 1981 they warned the other OPEC nations not to take it for granted that they would reduce to 7 mbd. As prices slumped and the 'Aramco advantage' became the Aramco albatross, they threatened dire consequences to any Aramco underlifter. The Saudis waited almost too long. Doubtless they expected demand to revive, but they may also not have taken due account of changes in market structure since 1978.

The OPEC nations sell crude oil to a motley crew of multinationals, independent refiners, governments, consumers, and traders. With the disappearance of the old integrated multinational channels, there is more buying and selling than ever. Organized futures markets have been successful in middle distillates. We have come a long way toward a wide decentralized market, which is much harder to watch. Hence the new OPEC committee appointed to do just that. The decontrol of US crude oil prices has put OPEC crude oils into daily competition with dozens of crude oils of all qualities. US crude prices have always adjusted quickly to small changes in supply and demand. It is a more important and more upsetting link than is the better known tie between North Sea and Nigerian crude, because the United States produces over four times as much as the North Sea.

Oil companies have learned to bargain, and walk away from renewing term contracts. A year ago, the Kuwait oil minister accused Gulf and BP—he must have blushed pronouncing it—of 'haggling'. Tradesmen, not gentlemen! Non-OPEC prices have been persistently lower than OPEC prices. Thus the OPEC nations become the suppliers of last resort.

In the first three months of 1982, the consensus in the oil trade was that the Saudis were trying to force down the price, or that the price somehow had to come down because of oversupply. Neither theory made sense. Short run, lower prices will not help sales. Better to sell the too few barrels at a higher than a lower price. All that matters is whether the group can hold together to limit output.

By late March, the OPEC nations had learned their lesson: the Saudis were not going to bail them out unless all agreed to limit output. OPEC drew up its first detailed prorationing scheme, limiting exports to 15.5 mbd. (Production of 17.5 plus 1 mbd natural gas liquids less 3 mbd local consumption.)

In April and May, demand continued to fall. Exports actually dropped another million barrels daily. Yet spot prices recovered dramatically. The accord sterilized nearly all the excess capacity, and assured each cartelist that his fellows were not about to undercut him by offering more supply. Iran has flouted the agreement; this would have been fatal when they had 7 mbd capacity; it did not much matter, now that they have perhaps a million in excess of their allotment. If others join in flouting the quotas, my guess is that the Saudis will swallow the affront, cut back, and look for countermoves.

Thus the supply-demand balance had no effect on prices; the success of the cartel was all that mattered. It is time the lesson were learned: that oil conservation and oil production in the consuming countries do not lower prices and do not help security, except to the extent that they weaken the cartel.

For consumers, the market is in a dangerous condition right now. The export mechanism has been wound very tight. If OPEC exports revive strongly, then another explosion is a good bet.

THE SURVIVAL OF THE CARTEL

I think the cartel will survive its problems, which are acute. Matters are difficult even at exports of 20 mbd. The non-Persian Gulf producers have over 10 mbd of capacity, are now using 5 mbd, and could hardly be kept from producing 7 mbd. That would leave 13 mbd for the whole Gulf. If peace broke out between Iran and Iraq, the production-price line probably could not be held.

Saudi Arabia will if need be use force to shut down or reduce its small neighbors' output. But the northeast is different. It was a good investment to pay $1 billion monthly to keep the Iran-Iraq war going. Now victorious Iran is demanding reparations to the extent of $150 billions or some such modest sum; the Saudis are apparently ready to pay up something. Surely, there is an Arab equivalent to Kipling's jibe about the trouble with paying blackmail: 'Once you have paid the Danegeld / You never get rid of the Dane'.

One way or another, the three large Persian Gulf producers could probably manage to divide 13 mbd among them. But if total OPEC exports were as low as the current 15 mbd, and the Persian Gulf were left with 8 mbd (i.e. 15 less 7 mbd), something would have to give.

Stagnant prices and exports strain their internal cohesion. OPEC export revenues went from $133 billion in 1978 to $275 billion in 1980, then declined to $255 billion in 1981. Assuming 1982 exports around 17 mbd, oil prices unchanged from last year, and an inflation rate of 10 percent per year, the real value of their 1982 revenues would be down 38 percent in two years, and be now only about 6 percent above 1978. Of course such calculations are very rough, but they suffice to let us know there is a budget battle inside each nation, which will exacerbate bargaining among nations. A spending squeeze does not make for greater willingness to assume burdens.

The best guess at the short run outlook is: Saudi Arabia must accept its responsibility to maintain prices, and must reduce its output enough to make way for the uncontrollable small OPEC members. They are not in a position, at this moment, to use force against Kuwait or the UAE; still less against Iraq or Iran. To teach their OPEC partners a lesson, by maintaining production and matching their lower prices, would risk a price collapse to unthinkable levels. They had better accept, with what grace they can muster, that the others are now returning the Saudis' favor of 1976–77, and 1979–82: cut price and gain additional sales. The Saudis will for a while need to live off their accumulated assets.

But in the long run, pushing Saudi Arabia into the classic position of the largest member, forcing it to curtail output more than the others, is only a second-best bad tactic for the cartel. For the elasticity of the particular demand curve facing the Saudis becomes drastically higher than that for the group as a whole—about 2.5 times as high, and increasing as its market share drops. Then as inflation gradually erodes the real price of oil, it may again be profitable for the whole group to raise the price, but not for the Saudis. At a higher price, there will be less consumption, and they will bear the whole brunt of the lost sales. The Saudis will refuse to raise prices without adequate assurance of a 'proper' market share for themselves. In other words, there will need to be some kind of market agreement of the type they reached in March 1982, and which is now rather attenuated, though not dead.

To be sure, revolution and war are a great help, and they are always likely around the Persian Gulf. Even a low level of sea-air activity, unskillfully carried out, can shut down loading of oil, as the Iran-Iraq war shows. But lacking such good luck or good management, it is hard to see the cartel able to maintain the current world price level, or a higher one, all by themselves. They will be forced, I think, to appeal to the governments of Norway, Britain and Mexico to join in output restriction to maintain prices, and to the United States for benevolent neutrality.

The United States may be receptive. It will be said that managed markets are stable, oil is too important to be left to the vagaries of competition, etc. The US State Department said, in rightly claiming credit for the 1971 Tehran agreements: 'the previously turbulent oil market would now settle down'. Mr Henry Kissinger was strong for a 'just price' when he was in office. In the 1980 Presidential campaign an agreement with the OPEC nations was suggested by Mr Kennedy on the left, Mr Connally on the right, Mr Carter in between. It was also endorsed at the 1980 summit. We may now have a super-Tehran as the first stage of 'global negotiations'.

On the other hand, the consuming countries may decide to divorce an inescapably high internal price level from the world oil price level, which would then sink to a small fraction of its current level. That is less likely, but not impossible.

A study of market structure explains why we will long have both an oil glut—since oil could be profitably produced in much greater amounts—and a chronic shortage, since the glut can only be warded off by curtailing output, and now and again over-curtailing it. The worldwide gas glut will persist until prices decline considerably in relation to oil and coal.

The least likely scenario is for oil and gas prices to drift upward to the end of the century. That would make sense if the market structure were competitive, and the price were in the neighborhood of long run marginal cost. Then prices would rise with costs. But because the price is so many times cost, changes in cost are of little importance. What matters is whether the cartel will flourish or fade. Natural gas availability will complicate its task.

Chapter 3

Towards a System of World Oil Regions: The New Geopolitics of International Energy

*Peter R. Odell**

The idea of global interdependence in energy is conventionally thought to be in the process of undergoing significant intensification with, of course, potential major consequences for the western world's international economic and political relationships. I want to hypothesize an alternative view, viz. that the internationalization of coal and gas will necessarily proceed relatively slowly with a consequential geographically rather limited impact. Even more important this slow process will be accompanied by a process of the de-internationalization of the global oil industry. The impact of the latter will be relatively more important than the internationalization of gas and coal, so that energy global interdependence will decline significantly over the period up to the end of the century and beyond.

The components underlying the hypothesis are:

1. A slow increase in the use of energy in general can now be expected so that the need for and the opportunities of expanding the markets for internationally traded coal and natural gas will

*Professor and Director, Rotterdam Centre for International Energy Studies, Erasmus University, Rotterdam.

remain much more limited than hitherto generally expected. In particular, it is the industrial countries which, for organizational and other reasons, would be best able to create an import potential for alternatives to oil, that will have the lowest energy growth rates. Third World countries, with higher energy growth rates are less likely—indeed, less well able—to change from their traditional high level of dependence on oil and so are not well placed to provide market opportunities even for low-cost gas and coal from distant resources.

2. There will be an even slower—or perhaps even a negative—growth in the non-communist world's use of oil. Oil is, indeed, rapidly becoming the energy of last resort in all markets where the OPEC controlled price has had to be accepted or has been voluntarily introduced (as in non-OPEC oil producing countries in which governments have not controlled the prices of locally produced oil). This process has already severely undermined the growth of international oil exports and is thus leading to the reversal of the earlier emphatic development of the internationalization of oil.
3. Concurrent with this demand side change in the outlook for oil, but achieving importance even more slowly, is the supply side response to the 12-fold rise since 1970 in the real price of internationally traded crude oil. This takes the form of an increasingly geographically dispersed pattern of oil exploration and exploitation. The slowness of this change is not only a function of the long lead times involved in establishing new oil production capacity, but is also a result of the reorganization of the industry which is needed to enable it to adjust its exploration and production efforts away from its hitherto near exclusive concern with the low cost, familiar and relatively low risk opportunities for oil exploitation in the OPEC countries and North America. Nevertheless, the process is under way, it is intensifying and it is gradually producing a more diversified geography of oil production. This means that a steadily increasing number of countries are, first, reducing their import requirements; and then, in many cases, achieving the prospect for self-sufficiency in oil. Though this is being done at costs which are higher, or even much higher, than the resource costs involved in the production of their oil requirements in the OPEC countries, the costs involved are still lower than the expenditures incurred to acquire oil imports from OPEC countries at the international prices which have been established over the last decade by the powerful combination of politico-economic forces involved.

THE IMPLICATIONS OF THIS PROCESS

The inexorable move towards a diffuse and dispersed pattern of oil production will produce a restructured non-communist world oil industry in which four components are of particular importance, viz.

1. Exports of oil from the traditional exporting countries will become the residual, rather than the central, element in the system.
2. Non-Middle East members of OPEC will find their best interests served by close association with and/or membership of regional organizations devoted to the establishment of high levels of energy self-sufficiency.
3. There will be increasingly intense competition between those countries with large potential export surpluses of energy—notably oil and natural gas—which are not linked into increasingly energy self-sufficient regions of the non-communist world. This applies especially to the Soviet Union with its economic need to export energy to the west and to the major oil exporters of the Gulf. The latter will be increasingly obliged to use their oil/gas production as low cost inputs to energy intensive industry as a way of maintaining export earnings.
4. The world-wide infrastructure and organization of the industry will become much more complex—in terms of physical elements such as transport and refining/processing facilities and the structure of those oil companies which decide to try to continue to operate internationally as oil companies rather than diversifying into other activities.

EMERGENCE OF A SET OF OIL REGIONS?

One likely result will be the emergence of a number of wholly or largely oil self-sufficient regions. Within each of these an increasingly integrated set of linkages will be developed between producing, refining, trans-shipping and consuming areas. Concurrently, the external connections of each region in respect of oil will be of decreasing importance—and eventually even of an intermittent nature. World trade in oil, and the international organization of the oil industry with which we have become so familiar, will be replaced by intraregional trade and a regional pattern of industry organization, respectively.

1. An East/South East Asia region

This is perhaps the clearest development to date. On the producing side Indonesia, Brunei, Malaysia and Thailand are already significant contributors or are poised to contribute more emphatically to the regional supply. Further geographical diversification of production can also be anticipated as a result of expanding exploration and development efforts in a region where oil and gas prospects are not only extensive, but also good. Organizationally, Singapore provides the fulcrum for the region and the already noticeable strengthening presence there of many oil companies can be expected to intensify. The region's relatively rapid rising demand for energy (a function of Japan's continued economic growth and the more recent industrialization of a number of other countries) provides the incentive for enhanced levels of production—particularly for Indonesia with its massive potential for natural gas production and an expanded oil producing industry. Guaranteed outlets in the region for its energy exports eventually seem likely to outweigh the importance of its membership of OPEC. For Japan, both guaranteed energy availability from relatively nearby producers, and the opportunities provided for its exports in the region, indicate a possible resurgence of Japan's earlier idea of a 'greater co-prosperity sphere' for this world region.

2. Middle America

This region's focus lies in the southern Caribbean in an area of the hitherto divisive rivalries of Venezuela and Mexico. The two countries have, however, already jointly accepted responsibilities for supplying oil to their poor neighbours on favourable terms, whilst their large long-term future export potential of conventional and/or unconventional oil provides a secure resource base for the whole of the region, including the potentially very large needs of Brazil as well as other smaller countries. These prospects, coupled with the increasing chances of tying the US into a western hemisphere energy supply system, suggest a net advantage for a regional commitment in place of continued OPEC membership for Venezuela and Ecuador—in cooperation with Mexico and other potentially larger oil (and gas) producers in the region. The region's organizational fulcrum is less evident than in the case of Singapore, but the pre-existing oil refining, trans-shipping trading and organizational functions of the Netherlands Antilles, together with their relative political stability suggest one possibility. This possibility is augmented by their status of neutrality between the interests of Venezuela, Mexico and the United States.

3. The 'Southern Round'

The 'Southern Round' comprises the countries of the southernmost parts of the continent of Latin America and Africa, together with Australasia. Oil and gas prospects in Argentina and on its extensive continental shelf, plus those of Australia and New Zealand, provide a basis for the region's self-sufficiency for many decades into the future, whilst common interests, arising from global geo-political considerations, will provide an increasingly powerful motivation for co-operation over energy supplies. The elimination of the need for imports from other parts of the world will thus undermine the impact of the embargo on oil supplies to some parts of this region by the member countries of OPEC so that the embargo now seems likely to be of little more than relatively short-term significance. The expansion of the region of the 'Southern Round' to take in the fringes of Antarctica, with its promise of hydrocarbons' potential, extends the scope of the region both geographically and temporally. Indeed, such a development could open up supply prospects and the need for markets outside the region by the turn of the century.

4. Western Europe and the Mediterranean Basin

A fourth region in prospect, though one which currently has little internal cohesion, comprises Western Europe and the Mediterranean basin. Here, the large, and potentially much larger, oil and gas producing countries of the North Sea basin and adjoining north-western European off-shore areas, plus the easily linked-in suppliers on the southern side of the Mediterranean Sea provide the reserves and the production potential which could eliminate the dependence of this most energy intensive using region of the world on supplies from other parts of the world. Refusals to license potential producing areas, hesitancy in allowing oil and gas which is discovered to be produced, and the propensity of many of the countries concerned to tax the oil companies into lethargy as far as their investment plans for the region are concerned, combine to produce a case-study on how not to approach resource development questions. Nevertheless, there are fundamentally more powerful forces which could come into play to create a potential for cohesion over energy developments. These include, first, the growing impact of an increasingly hostile external world in general, and threats from the major powers, in particular, to the security and/or welfare of the region: and, second, the recognition by countries as diverse in their current interests and policies as Algeria, Libya, the United Kingdom and Norway that guaranteed markets in Western Europe for their oil and gas are preferable to the uncertain future which faces their indigenous oil and gas industries as a result of the changing international energy/oil situation.

Changes in policies consequent upon the influence of these factors could produce a region potentially self-sufficient in oil and gas production. Moreover, most of the refining, transport and distribution infrastructure required to get the quantities needed to the markets is already in place, or it can be easily and quickly developed. External relationships in respect of the future energy import needs of the Western European/Mediterranean region, such as they will then need to be, will be to the East. From there the Soviet Union, on the one hand, and countries of the Gulf on the other, will have to compete for the very limited markets which are likely to be available.

5. Other Regional Developments

The oil regions discussed above by no means cover the world. Other regions are possible—including West/Central African and circum-Indian Ocean regions—but there are still insufficient indications of their development to justify even initial speculation about them. China is also excluded. It could be the producing centre of an East Asian region with links to its neighbours to the south and, even more important, to Japan; but a more likely prospect is its integration into the East/South West Asian region already described. On the other side of the Pacific, a region comprising Canada plus the United States is an alternative to the latter's involvement with the Middle America region, providing Canada does not pursue its isolationist oil/energy policy for too much longer.

MAIN ACTORS IN THE CURRENT GLOBAL OIL SYSTEM

The position of the Soviet Union, the oil producing states of the Gulf and the international oil companies is now considered.

1. The Soviet Union

The Soviet Union appears to be wealthy enough in its potential resources to be able to sustain the future oil and gas needs of the Eastern block for the rest of the century. Given positive co-existence with the United States, leading to technological and financial help for the more effective and rapid development of the Soviet oil and gas potential, it would also be able to expand its present commitment to supply energy to Western Europe. As shown earlier in the paper, however, a cohesive Western Europe/Mediterranean region will have limited needs of oil and gas imports. In this context the Soviet Union would have to compete against the other contemporary major oil region of the world that is excluded from this analysis; namely, the Gulf, the heart-land of the international oil system over the last 30 years.

2. The Oil Producing Countries of the Gulf

One could argue that these countries have created conditions which necessarily exclude them from the future world of oil; the consequences of policies which have pushed the price of their oil exports so far above the long-term supply price of the commodity as to make their production decreasingly relevant in a world which, with a much lower demand growth rate, has alternative oil supply options open to it. This is especially true when the economic element in the argument is coupled with the fears for the security of the oil supply potential of the region as a result of issues such as the Arab/Israeli dispute, the conflicts and potential conflicts between contiguous oil rich countries of the region, and the rise of Islam militant.

The significance of the potential for a possible relatively near-future reshaping of the world oil industry along the lines indicated in this paper will not, however, be lost on the Gulf producers. They may thus be expected to try to contain the developments suggested. They could, for example, use the power they are able to exercise over nations currently dependent on oil imports from the Gulf to persuade them not to seek regional agreements for self-sufficiency in oil. Or, they might attempt to 'buy' their way into the regionalizing system, with Western Europe as their most likely regional objective in this respect. Failure would diminish the importance of the Gulf not only in world oil terms, but also more generally in world geopolitical terms. Unless the Gulf oil producing states decided that their interests were likely to be best served by a close alliance with the USSR in order to achieve an agreement on a market sharing arrangement for their limited oil and gas export opportunities. This would be a highly paradoxical development, given current fears in some circles over Soviet plans to take control over the Gulf region in order, either to ensure its own oil supplies, or to deny oil to the west. It would certainly be a geo-political change with immense implications and the possibility, in itself, may suggest to the western industrialized world that its oil imports from the Gulf ought to be maintained at levels higher than those which would remain, were the process of the readjustment of the world of oil into wholly or largely self-sufficient regions to proceed unhindered, on economic and geo-political grounds.

THE INTERNATIONAL OIL COMPANIES

Finally, it is clear that the international oil companies will also need to rethink their global strategies for continued profitable operation in the changed world oil system as hypothesized. They will need to reshape their own organizational and control mechanisms in order to reflect the newly emerging regional pattern of the late 20th century oil industry.

Some of the companies seem likely to be more flexible than others in respect of these needs and thus be able to retain an element of managerial and technological internationality in the context of the regionalized oil system. Some of the companies, on the other hand. may decide to pursue a policy which restricts their oil and other activities to the United States in order to ensure what profits they can in an organizationally much less demanding 'fortress America'. However, the impact there of the long held belief in, and the likely enhanced efficacy of, government controls over oil companies (the present Administration excepted), as well as the absence in the US of inherently low cost oil and gas resources, may make it more difficult for oil companies to continue to earn a respectable living in the US, compared with the greater—though more risky—chances for profitable operation in the more challenging world outside, where most opportunities for oil and gas developments on a large scale remain largely untapped.

Chapter 4

Changing Markets for Coal: Opportunities for the Developing Countries

Rajendra K. Pachauri and Walter C. Labys**

While the consumption of coal in the past has been concentrated in the nations of America, Europe and the communist world, major economic changes are taking place in the developing countries which warrant a closer look at the position of coal for energy consumption in these nations.[1]

The present moment is particularly appropriate for an analysis of coal use in the LDCs, since the immediate period after the 1973-74 oil price rise was marked by much wishful thinking on the future of coal ('the United States would emerge as the OPEC for coal,' etc.) which percolated through the literature. On the other hand, more recently, the oil glut of 1981-82 and the relative lull in oil prices appear to have blunted the urgency of a rapid transition to coal, almost universally advocated earlier. The middle ground between

*Dr Pachauri is now at the Tata Energy Research Institute, Bombay and Dr Labys is with the Massachusetts Institute of Technology Energy Laboratory, Cambridge, Mass. This paper was written when the first mentioned author was Visiting Professor at West Virginia University, Morgantown, W. Virginia, and later Research Fellow, East-West Resource Systems Institute, Honolulu, Hawaii. The authors would like to thank B. C. Kim and U. Soelistijo for help in preparing this paper. The full text including the case studies of India, Indonesia, South Korea and Colombia is lodged and available in the BIEE Archive in Chatham House, London and IAEE Archive in Washington.

these two extreme positions requires careful mapping and analysis in the light of current trends and future plans.

ANALYSIS OF PAST TRENDS

Coal production, consumption* and trade have grown significantly during the seventies. Total world production of coal in 1980 included 2759.8 million metric tons (mmt) of hard coal and 960.8 mmt of lower ranks. Of these quantities, production was concentrated in only eight countries, with the USA producing 710.3 and 42.3 mmt respectively, the Soviet Union 495.0 and 162.90 mmt, China 620.0 mmt of hard coal only, Poland 193.0 and 36.9 mmt of hard coal and lower ranks, the United Kingdom, South Africa, Australia, and India producing 128.2 and 0 mmt, 115.0 and 0 mmt, 84.3 and 32.9 mmt, and 109.0 and 4.5 mmt respectively.[2] The total resources of coal in the LDCs and their production in 1980 are shown in Table 1.

It can be seen that the largest share of world coal resources lies outside the Third World, and, therefore, growth in consumption has direct implications for growth in international trade in coal. Yet there are countries, particularly in Africa, which hardly produce any coal, despite the availability of significant coal resources. Among the developing regions of the world, Asia produces the largest share, but this performance is dominated by production in China, followed by India. There is, therefore, potential for increased production of coal in the LDCs, particularly in Africa, and in other regions including countries such as Mexico, Chile, Colombia, Venezuela, Indonesia, and Turkey. Whereas these increases in production would contribute significantly towards meeting LDC demand for coal they would hardly be large enough to reduce the importance of coal imports to the Third World, and coal trade is likely to grow substantially in the future. The levels of international trade in coal for the years 1970 and 1980 are shown in Table 2. It would be seen that in the decade 1970–80 coal exports and imports have grown by 49 percent.

There are some specific features characterizing international trade in coal which can be summarized as follows:

1. Coal exports are dominated by a small group of countries. In 1980, only seven countries accounted for 94 percent of total exports (as against the case of OPEC where in 1981 the largest

*See Statistical Annex, Table M.

Table 4.1. Coal Resources and Production in the LDCs, 1980 (mmt)

Country	*Total resources all ranks*	*Percent of world resources*	*1980 Production*	
			Hard coal	*Lower ranks*
North and Central America				
Mexico	3,280	<0.1	3.2	–
South America				
Brazil	15,807	0.1	5.1	–
Chile	4,426	<0.1	0.8	–
Colombia	10,063	<0.1	4.9	–
Venezuela	9,178	<0.1	–	–
Other	10,748	<0.1	1.1	–
Africa				
Botswana	107,000	0.8	–	–
South Africa	92,511	0.7	115.0	–
Morocco	140	<0.1	–	–
Mozambique	425	<0.1	–	–
Swaziland	5,020	<0.1	–	–
Zaire	NA	NA	–	–
Zambia	130	<0.1	–	–
Zimbabwe	8,310	<0.1	3.1	–
Other	4,361	<0.1	2.5	–
Asia				
India	114,034	0.8	109.0	4.5
Indonesia	10,117.6	0.15	–	–
Bangladesh	NA	NA	–	–
Pakistan	646	<0.1	1.5	–
South Korea	1,231	<0.1	18.6	–
Thailand	NA	NA	–	1.5
Turkey	5,412.7	<0.1	4.3	16.2
TOTAL WORLD	13,609,298	100.0	2,759.8	960.8

Source: George Markon, 'World Coal Trade in 1980', *World Coal* 7 (No. 7), pp. 43–49 (Nov./Dec. 1981).

seven exporters handled only 85.6 percent of the non-communist world's total exports).

2. If we exclude the communist nations, with the exception of South Africa, all the major exporters are developed nations. None of the developing nations has large enough known resources to be able to emerge as a major exporter.
3. The growth of the coal industry in the major exporting nations has been historically around domestic markets, unlike the major

Table 4.2. World Coal Trade, 1970-1980: Major Exporters and LDC Importers

Country	*1970*	*1980*	*Change 1970-80 (Percent)*
Exporters			
United States	65,075	83,199	27.9
Canada	3,987	14,311	258.9
West Germany	15,906	12,661	−20.4
France	1,213	468	−61.4
Belgium	590	563	−4.6
Netherlands	1,646	1,548	−6.0
United Kingdom	3,191	1,134	−64.5
USSR	24,499	24,500	0.004
Poland	28,816	30,949	7.4
Czechoslovakia	2,970	4,000	34.7
Australia	17,965	42,820	138.4
South Africa	1,507	28,464	1,788.8
PR of China	227	6,200	2,631.3
Other	1,701	1,588	−6.6
TOTAL	169,293	252,405	49.1
Importers			
America			
Argentina	746	922	23.6
Brazil	1,987	3,997	101.2
Chile	236	250	5.9
Mexico	151	609	303.3
Asia			
South Korea	17	7,296	42,817.6
Taiwan	—	4,069	
All others	1,067	4,674	338.1

Source: World Coal 7 (No. 7), p. 50 (Nov./Dec. 1981).

oil exporting nations where production of oil was essentially first developed to satisfy demand in other nations. Australia and South Africa have, however, developed large export capacities relative to domestic demand.

4. The major growth in demand for imported coal has taken place in Japan and the countries of Europe and Japan. Demand for coal imports in the LDCs has been relatively slow in growing. Of the various regions of the world, LDC demand has grown most rapidly in East and Southeast Asia.

Given the limited potential for increased coal production in the LDCs, changes in the demand for coal in the Third World would

Table 4.3. Coal Imports by Region (MTCE)

	1977 (Actual)	Projected Imports 1985 Case A	Projected Imports 1985 Case B	Projected Imports 2000 Case A	Projected Imports 2000 Case B
Steam Coal					
East and Other Asia	—	5	24	60	179
Africa and Latin America	1	3	3	6	10
Metallurgical Coal					
East and Other Asia	3	10	16	40	48
Africa and Latin America	7	20	20	57	80
TOTAL	11	38	63	163	317

Source: Report of the World Coal Study, Tables 2-4, and 2-5.

bring about corresponding changes in the international trade in coal. At the same time, the perception of future availability of coal imports would continue to be a major determinant of decisions which directly affect the demand for coal.

PROSPECTS FOR INTERNATIONAL TRADE IN COAL

Various studies in recent years have addressed the question of coal use for energy applications in the LDCs and forecasts have been made for international trade in coal. The World Coal Study (WOCOL)[3] forecasts for coal imports in the LDCs are shown in Table 3.

The largest growth in demand, according to these projections, would take place in East and other Asia. But the variability of these projections is brought out by the wide range between WOCOL's two scenarios for coal imports in the year 2000. The major share of this demand will be generated by the countries of East and Southeast Asia, and would come from a major expansion of existing coal-based power capacity. According to the World Bank[4] coal/lignite-based power capacity in the LDCs would grow from 35.1 Gigawatts in 1980 to 92.2 Gigawatts in 1990. Countries contributing to this growth include the Republic of Korea, Taiwan, Philippines, Thailand, Uruguay, Panama, etc.

The International Energy Agency has studied the global prospects

Table 4.4. Projected World Coal Trade (Scenarios A and B)

	1985		*1990*		*2000*	
	A	*B*	*A*	*B*	*A*	*B*
Canada	−6	−6	−9	−12	−14	−29
United States	−68	−68	−79	−81	−129	−214
OECD/Europe	+94	+116	+153	+179	+311	+468
Japan	+96	+101	+123	+142	+181	+231
Australia/New Zealand	−57	−82	−90	−136	−195	−259
Total OECD	+59	+61	+98	+92	+154	+197
Centrally Planned Economies	−43	−43	−49	−44	−66	−66
Developing Countries	+11	+10	+2	+2	−10	−56
South Africa	−34	−34	−60	−58	−90	−90
Others	+7	+6	+9	+8	+12	+9
Total World Coal Movement	208	233	287	331	504	708

Source: Ulf Lantzke, 'Expanding World Use of Coal', *Foreign Affairs.*

for coal as a source of energy, and their projections are presented and discussed in a recent paper by Lantzke.[5] Projected levels of world coal trade under two sets of conditions are presented by Lantzke in terms of a scenario (say A) assuming present policies and another scenario (say B) assuming pro-coal policies. These are shown in Table 4.

Irrespective of the differences shown in scenarios A and B, and questions on the accuracy of the forecasts shown in Table 4, some important conclusions can be drawn from the pattern of supply and demand expected in the future. These are broadly as follows:

1. In the non-communist world, the major sources of exports in the future would be the USA and Australia, with significant contributions in supply from South Africa.
2. The IEA has downplayed the importance of coal in the LDCs and has projected very low levels of import demand in this group of countries which are not borne out by other studies; demand would be partly met by increased domestic production (India, Colombia, Mozambique, Venezuela) and substantial exports from China, leaving a residual demand of 10 mmt and 56 mmt in 2000 under the two scenarios.
3. Based on the geographical distribution of demand, which is concentrated mainly in East and Southeast Asia, the growth of Australian coal supplies for this segment of the international market assumes considerable importance. Exports from the

western part of the United States can contribute significantly to meeting future growth in demand in the Pacific rim region, but the major constraints in development of western coal in the United States leave these prospects in considerable doubt.

CONCLUSIONS

The country reviews we have undertaken illustrate the complexities and constraints governing a shift to greater use of coal in the developing countries. While there is a strong commitment to expand coal production and consumption in many nations, coal markets in the LDCs may develop slowly and, perhaps, far below desired rates in the immediate future. The potential for expansion of coal use is, however, very great, and if successfully tapped this could result in an easing of pressure on the global oil market by the end of the eighties—it needs to be emphasized that the demand for oil is growing much more rapidly in the LDCs than in the OECD countries.

Major increases in coal consumption will take place in Asia, but this region would be sensitive to the uncertainties of supply from the United States, in particular and, to a lesser extent, from Australia. Apart from serious problems in expansion of western coal in the United States, there is always the relevant question of priorities which would always be resolved in favor of US exports to Europe and Japan in preference to Asian LDCs. Even though Australia's efforts at expansion of coal exports appear more tangible than those of the United States, transport and port handling constraints and labor problems may limit, both, quantities and reliability of Australian coal exports. In the absence of assured future supplies, the developing countries of Asia may not invest large resources in coal-dependent technologies and equipment.

REFERENCES

1. See for instance, 'Report of the World Coal Study', Carroll L. Wilson, Director, *Coal—Bridge to the Future*, Ballinger, Cambridge, MA, 1980; and World Bank, *Coal Development Potential and Prospects in the Developing Countries*, Washington, DC, 1979.
2. *World Coal* 7 (No. 7), p. 50 (Nov./Dec. 1981).
3. Report of the World Coal Study, Carroll L. Wilson, Director, *Coal—Bridge to the Future*, Ballinger, Cambridge, MA, 1980, Tables 2-4 and 2-5.
4. World Bank, *Energy in the Developing Countries*, Washington, DC, 1980, p. 45.
5. Ulf Lantzke, 'Expanding World Use of Coal', *Foreign Affairs* 58 (No. 2), pp. 351-373 (Winter 1979/80).

[illegible] part of the United States [illegible] affords [illegible]
[illegible] growth in demand in the Pacific [illegible] region
[illegible] major constraint in development of western coal [illegible]
[illegible] in considerable [illegible]

CONCLUSIONS

The country reviews we have made [illegible] the com[illegible] [illegible] general [illegible] to greater use of coal in the developed [illegible] there is a strong [illegible] to expand coal [illegible] and [illegible] national [illegible] coal [illegible] [illegible] desired [illegible] in the [illegible] future. The potential for expansion of coal use [illegible] [illegible] the [illegible] [illegible] [illegible] of the eighties [illegible] grow as much [illegible] the OECD countries.

[illegible] place in Asia, [illegible] of supply from the United States [illegible] development, [illegible] of [illegible] coal [illegible] the United States [illegible] in Europe and Japan [illegible] efforts [illegible] the United States [illegible] and labor [illegible] of [illegible] developing countries [illegible] coal dependent [illegible] and equipment.

REFERENCES

[illegible]

Chapter 5

Recent Developments in Nuclear Energy: A West German Perspective

*Fritz Lücke**

In my opinion, we have come to a crossroad in nuclear energy, where on the one hand serious risks are looming, but also on the other hand, the cost and environmental advantages of nuclear energy are becoming more obvious to a broader public.

Although the world-wide role of nuclear energy has been steadily increasing over the last couple of years its fate has been quite different in individual countries.

- France and the United Kingdom are outstanding examples of a steady and continuous expansion of nuclear energy without any very major political problems.
- In Austria, Switzerland and Sweden, on the contrary, nuclear energy has become a major political issue on which elections and referenda have been based.
- In the United States nuclear energy had to face a particularly difficult situation since the Three Mile Island accident, not only for psychological reasons, but also for cost considerations, due to costly additional security requirements and the competition of the increased use of coal. It will still have to be seen, if the new energy policy implemented by the Reagan Administration can bring forth a fundamental change.

*Landeswirtschaftsministerium, Kiel.

THE POSITION IN WEST GERMANY

The Federal Republic of Germany is probably the country where emotions and ideology have most influenced the discussions about the role of nuclear energy. Although there are still more people ready to accept the necessity of nuclear energy for the long-termed securement of our energy supply than there are against it, the number of opponents has been steadily increasing until very recently. On the other hand it is interesting to know, that only a very small minority is really prepared to accept price increases for their energy use in order to enable the development of alternative sources of energy or the increased use of coal in electricity generation.

This ambiguous situation, further enhanced by a lack of political leadership, has been reflected both in the actual development of nuclear energy in the Federal Republic as well as in the nuclear energy policy of the last few years. This becomes quite evident by the following facts.

In my region, Schleswig-Holstein, the nuclear power plant of Brokdorf has become something of a symbol for the anti-nuclear movement. Battles have been fought around the construction site, which, at that time, was not more than a meadow and a fence. Now the construction phase has been entered. When this power plant, according to the actual planning, will be commissioned in 1987, its construction costs will have doubled (to somewhere between 3 and 3.5 Billion Deutschmarks) and its construction time will have extended to more than 13 years.

In general, construction of nuclear power plants in the Federal Republic of Germany has significantly slowed down. It was only in December 1981, that for the first time for three years a new nuclear power plant (Grafenrheinfeld) was plugged into the electricity grid. Since more than three years no new applications for the construction of nuclear power plants have been submitted to the authorities.

Projections and scenarios for the role of nuclear energy in our future overall energy supply had to be revised accordingly. Nuclear energy's share in our total primary energy requirements in 1981 was 4.6 percent relating to a share in total electricity generation of about 14 percent. The independent institutes' projections for 1995, which were computed in preparation for the third revision of the energy policy programme of the government of the Federal Republic of Germany in spring 1981, foresee nuclear energy's share increasing to some 17 percent of TPE which represents some 37 Gigawatt of generating capacity. Only four years ago the second revision was based on a prediction of 40 Gigawatt for 1990. The first revision of November 1974 predicted 45 Gigawatt already in 1985. But even the

1981 forecast appears optimistic, since it implies the necessity to construct 12 additional nuclear power plants of the 1300 Megawatt type by 1995. It is questionable whether this objective can be achieved after the experiences of the past.

But the controversy about nuclear energy has also had its direct impact on the political situation in the Federal Republic.

- In both ruling coalition parties there are strong anti-nuclear currents.
- This division is particularly significant in the bigger of the two coalition parties, where some of the regional congregations clearly oppose the official government energy policy. It was only after intensive discussions at this party's annual convention in Munich, that the request for a moratorium for the further construction of nuclear power plants could be avoided.

This political stalemate is most significantly reflected in the recommendations of the Enquiry Commission 'Future Nuclear Energy Policy' set up by the previous German Bundestag, which has identified four possible energy paths and recommends leaving the options open until the end of this decade; that chosen will determine the future role of nuclear energy.

The Federal Government is aware of the issues at stake and has, very much in response to the insistence of the Bundesländer, put forward a project to accelerate the nuclear licencing procedures. However, given the political situation, there appears to be little leeway left for political actions.

MAJOR AREAS OF DEBATE

At present three major areas of discussion can be identified: these are the standardization of nuclear power plants, the fate of the advanced reactor technologies and the timely implementation of adequate nuclear waste disposal facilities.

The standardization of nuclear power plants is one of the measures put forward in the above mentioned project in order to accelerate the licencing procedures for the construction of these plants and to constrain the rapid increase of construction costs. However, only a couple of months after the Ministry of Interior as supreme licencing authority had approved this new procedure, the same Ministry requested the implementation of a number of additional security requirements. This, according to the utilities involved, would have increased the costs by some extra hundred millions Deutschmarks. In a compromise between the Federal Government, the Bundesländer and the Utilities, this dispute was straightened out. The psychological damage, however, is considerable and it is still open, whether the three plants in question will be launched within the scheduled timetable.

This issue is part of a more fundamental discussion at present under way in the Federal Republic. It focusses on the question whether the official philosophy—that security has absolute priority over economic considerations and this has considerably contributed to the high security standards of nuclear power generation—is not on the verge of being overstretched. A number of serious critics claim that the present understanding of this philosophy, which lacks any definition of socially tolerable minimum standards, has led to gigantic red tape, an enormous inflation of costs and only insignificant increase in incremental security, if any. It appears a justified apprehension that nuclear energy could be at stake if this development is not brought to a halt and licencing procedures are not deflated and re-streamlined. In fact, Gottlieb Benz or Rudolf Diesel would not have been able to develop their epochal inventions under the present security philosophy.

The question of nuclear waste disposal has gained particular importance due to the fact that the principles of nuclear waste disposal, as defined by the Federal and Länder Governments on 29th February 1980, require that utilities prove that the necessary waste disposal precautions for each individual nuclear power plant have actually been taken. Although for the reactors actually in production or under construction this proof has been met, e.g. by means of reprocessing contracts with France and the United Kingdom, this linkage could impede the construction of further power plants, if progress to solve the waste disposal question would be insufficient. In the meantime, work for two intermediate storage facilities is well under way and considerable efforts are made for the identification of final storage sites. This will be in the salt-domes of Gorleben and the former iron-ore mine 'Konrad' near Saltzgitter. At least for the moment this encouraging progress has silenced those critical voices, which requested a halt to all additional nuclear power plants until the waste disposal question had been solved.

Considerable efforts are also under way to locate a site for a first prototype reprocessing plant. Relevant studies are most advanced in Hessia and in Bavaria. However, it is important that the actual momentum should not be lost, be it for political or for financial reasons.

By the mid-eighties, the question will also have to be decided by the Government as to whether a nuclear waste disposal system without reprocessing offers decisive advantages in terms of safety as compared with a reprocessing scheme. Research projects and studies on the subject are under way. A number of Länder representatives, including my own, have, however, already pointed out, that they would not consider waste disposal without reprocessing a valid option.

COST UNCERTAINTIES

The issue of the advanced reactor technology represents most significantly the dilemma of the present nuclear energy policy in the Federal Republic. Both the fast breeder reactor and also the thorium high temperature reactor have experienced excessive construction delays and cost increases. Due to very complicated and time-consuming licencing procedures, continuously updated security requirements and extensive, if not over-extensive technical tests and documents of every construction detail, both lines have by far exceeded the scheduled time and cost schedule and run into serious financial problems. Construction of the fast breeder reactor was started in 1972 at envisaged total costs of 1.5 Billion Deutschmarks. Today cost estimates are approaching 6 Billion Deutschmarks. Construction of the thorium high temperature reactor was started in 1971 at a cost estimate of 700 Million Deutschmarks to be completed in 1979. Today the total costs are estimated in the range of 3.7 Billion Deutschmarks and completion expected not before 1985.

Without entering into any details of this development, it is evident that this cost explosion and the time-delay has run those projects into serious trouble. The Government has claimed that the interested industry, in particular the utilities, would have to contribute a bigger share of financing, and in lengthy negotiations another 900 Million Deutschmarks have been permitted for the fast breeder reactor. Today additional financing for both reactor lines in the range of some 2 Billion Deutschmarks is required. Just as an order of magnitude let me mention that the French Super Phenix fast breeder reactor with a capacity of 1.200 Megawatt will be completed in 1984 and will then have cost as much as the much smaller fast breeder reactor in Kalkar.

The Government will take a final decision on this intricate problem at the beginning of October 1982. So far the result of the negotiations and deliberations under way is completely open, although all political parties have expressed their support for the continuation of both reactor lines. Whatever the result will be, it is quite evident that the cancellation of one line would have far reaching effects on the other line and that such a decision would do long-lasting damage to the role of nuclear energy in the Federal Republic.

PART III

The Changing Parameters of Demand

Abstracts (see Section 5)

Chapter 6

Energy Efficiency and Production Efficiency: Some Thoughts Based on American Experience

*Sam H. Schurr**

Although specific manifestations will change, there is reason to believe that the energy-technology-productivity nexus which has been so important in the past could exercise a similar—and perhaps even a greater—leverage on future productive efficiency and economic growth, and indirectly on energy productivity, as well.

The book, *Energy in the American Economy*[1] published in 1960, included a detailed quantitative analysis of the history of energy consumption in the United States between 1880 and 1955. One of the most interesting findings was that the relationship between the consumption of energy (measured in Btus of the basic mineral fuels and waterpower consumed) and the growth of GNP in the United States had been characterized by two distinctly different long-term trends, separated roughly by the decade of the first World War.

*Deputy Director, Energy Study Center, Electric Power Research Institute, and first recipient of the IAEE annual award for contributions to the literature of energy economics and for service to the profession.

†I am grateful to the following for reading an earlier draft and making useful comments: Charles A. Berg, Ernst R. Berndt, Calvin C. Burwell, Warren D. Devine, Jr., Walter Esselman, Solomon Fabricant, Charles J. Hitch, Milton F. Searl, Sidney Sonenblum, and Chauncey Starr.

The fact that the consumption of energy relative to GNP (i.e. energy intensity) had been declining persistently over a long period of time following World War I had been documented for the first time, to my knowledge, in a pioneering study by Harold J. Barnett published in 1950.[2] Our findings agreed with Barnett's results, but, by extending the historical analysis back to 1880, we discovered also that during an earlier period in American history the trend of energy consumption (mineral fuels and hydropower) relative to GNP had been persistently upward, in contrast to the (then) more recent trend of declining intensity.[3]

The existence of these two strongly contrasting periods came as a surprise to us, and we pondered the factors that could explain such a pattern. The part of our explanation that I want to discuss here focuses on the period of declining energy intensity, or, equivalently, rising 'energy productivity'. This decline occurred during the same period of time as the efficiency of the American economy, as measured by labor and total factor productivity, was improving at an exceptionally fast rate. The post-World War I period which marked the turning point in energy intensity trends also witnessed a distinct acceleration of national trends in the growth of labor and capital productivity.[4] *Thereafter, energy productivity, and labor and total factor productivity were generally all growing at the same time.*

Our initial hypothesis as to why energy intensity fell centered on two sets of factors that seemed to provide a *prima facie* explanation: (a) the changing composition of national output towards lighter industries and services which, on average, used less energy per unit of output than the heavy industries that had been more dominant in the past; and (b) major improvements in the thermal efficiency of energy conversion (i.e. more output of useful energy per unit of raw energy inputs) particularly within such large uses of raw energy as electrical generation and railroad transportation. These factors turned out to be important, but they provided only a partial explanation of the change in trend that occurred following World War I.

We therefore turned our attention to another set of factors that, it seemed, might have played a less obvious but very significant part: to wit, those factors associated with changes in the composition of energy supply, particularly the growing importance within total energy supply of such highly flexible energy forms as electricity and fluid fuels (petroleum and natural gas) as opposed to the direct use of solid fuels such as coal. We paid particular attention to the possible impacts of the expanded use of electricity since electricity had grown by a factor of about 10 from 1920 to 1955 while all other energy consumption had barely doubled.

A point that was of particular interest to us pertained to the distinction between the large thermal losses in generating electricity, on the one hand, and what we called the 'economic efficiency' of electricity use, on the other. Even though there had been sharp increases over time in the efficiency of converting fuels into electricity—measured by a decline from almost 7 pounds of coal equivalent consumed per electric kilowatthour generated in 1900, to about 3 pounds in 1920 and less than 1 pound in the mid-1950s—it still took (and now takes) several Btus of fuel to produce 1 Btu in the form of electricity. Electricity generation thus entails obvious thermal losses. However, it is also true that the unusual quality characteristics of energy in the form of electricity had helped to raise overall efficiency by making it possible to perform tasks in altogether different ways than if fuels had to be used directly, and this is what caught our attention.

The impact of electrification on industrial processes was a clear case in point. Electric motors as a percent of total horsepower in manufacturing grew from about 25% in 1910 to about 90% in 1939. And with the use of electric motors, which could be flexibly mounted on individual machines, the sequence and layout of productive operations within the factory could be made to match the underlying logic of the productive process, as opposed to the more constrained organization of production imposed by a system of shafts and belting linked to a single prime mover. According to contemporary accounts in industrial journals, the use of electrification within factories to overcome the rigid constraints imposed by mechanical systems of energy distribution was a factor of enormous importance in the growth of productive efficiency in manufacturing.[5]

What this told us was that the use of electrically-based machine processes supported increases in the growth of overall productive efficiency.[6] This was, of course, enormously important, but we carried the analysis one step further, and asked whether this development could also have led to a significant, indirect impact on improvements in the efficiency of energy use itself, despite the thermal losses sustained in the generation of electric power. We reasoned that because its positive effects on the overall efficiency of industrial processes yielded greater manufacturing output in relation to combined labor and capital inputs, electrification could also have led, in a roundabout way, to a decline in the amount of energy required per unit of (expanded) output. If this were the case, not only would electrification have supported the growth of labor and total factor productivity (more precisely, 'multi-factor productivity'), but it also could have enhanced the productivity of energy use.

We did not examine the impact of fluid fuels to the same extent, but we noted that the utilization of the internal combustion engine had played a similar role in permitting the substantial mechanization of agriculture which played so great a part in the rising productivity of the American economy and in releasing farm labor to other sectors of production. By extension, the same reasoning could be applied to the growth of truck transportation. This development made it possible for industry to move away from sites previously dictated by the presence of railroad and harbor facilities to locations more advantageous in other respects, thereby opening up new possibilities for improvements in the efficiency of production and transportation operations. As in manufacturing, this could also have led to a decline in energy intensity in relation to agricultural production, transportation services, and the output of other productive sectors, or, to put it the other way around, energy productivity could have risen.

MORE RECENT LESSONS

Let me try now to buttress and update the foregoing account with some statistics drawn from a more recent report.[7]

In the fifty years 1920–1969, with the single exception of 1953–1960, increases in total factor productivity and declines in the intensity of energy use occurred in tandem. What is equally striking is that, between 1920 and 1953, the periods of greatest growth in total factor productivity were usually also the periods of greatest decline in energy intensity. For example, between 1920 and 1929, total factor productivity in the industrial sector rose at a very high average annual rate of 5.1%, while energy intensity in that sector declined at the very high average rate of almost 4% per year. Similarly, between 1948 and 1953, total factor productivity in the economy grew at a brisk 3.4% on average, while energy intensity fell at a high average rate of over 3% per year. Although the inverse relationship between total factor productivity and energy intensity virtually disappeared during the 1953–1969 period, it is still noteworthy that high rates of improvement in total factor productivity were essentially not associated with increases in energy intensity.

Results such as these are striking and, at first blush, puzzling because they imply relationships between energy and other productive factors that appear to be counter-intuitive. Didn't the rise in overall productivity require the substitution of energy and machines for labor? And didn't the relative use of energy rise as a result? The answer is yes to both questions. Over the entire period 1920–1969, energy use was growing more than three times as fast as worker-hours employed, so that energy use rose *relative to labor inputs.* Therefore, wouldn't the energy intensity of production necessarily increase? The answer is no. The apparent reason is that as a result of the leverage on overall productive efficiency exercised by the new technologies (with electrically-based industrial technologies playing an important part), final output grew faster than energy consumption. The net result was that energy consumption relative to total national output of goods and services declined persistently. What appears to have been at work was a set of energy-technology-productivity connections which yielded an augmentation in total output large enough to more than offset the increased use of energy relative to labor.

* * *

When *Energy in the American Economy* was written, almost no policy concern existed anywhere for what is now called 'energy productivity'. If the idea had occurred to anyone that enhanced energy efficiency might become a major policy goal, this fact had not come to our attention. But now that it has become a major objective of national policy, I would like to stand back and try to restate the main threads of our analysis as they bear on today's situation.

Our explanation of the long period of declining energy intensity involves, as one of its main features, the postulated existence of a reinforcing set of relationships between energy and the economy that goes something like this:

1. Energy was not only cheap and abundantly available but increasingly in forms that were unusually flexible (i.e. electricity and fluid fuels) compared to the direct use of solid fuels that had previously dominated energy supply.
2. These characteristics of energy supply—low cost, abundance, and enhanced flexibility in use—led to the utilization of energy in a variety of new processes and in new industrial locations.
3. The direct and most important effect of these imaginative new applications was to quicken the pace of technical advance and this showed up in improvements in the efficiency of productive operations, as reflected in the behavior of such statistical indicators as labor productivity and total factor productivity.
4. The increase in final output resulting from the growth in productive efficiency was greater than the increase in energy consumption associated with production.
5. As a result, in a roundabout way that was unintended (and indeed unnoticed), the positive effects on overall economic output resulting from the aforementioned factors served also to enhance the productivity of energy, as measured by the relationship between the growth of national output and inputs of energy.

The net result, then, was that strong improvements in both energy productivity and overall productive efficiency were achieved without any special efforts being made to bring about this desirable combination of circumstances. Energy was abundantly available, and its price was low and, for the most part, falling during this period. Simple economic reasoning would tell us that the intensity of energy use should have risen because favorable energy prices would have encouraged energy consumption. But even though energy use rose relative to labor inputs it fell in relationship to the final output of the economy. Did this decline in energy intensity take place in spite of low energy prices, or somehow because of them? We don't really know for sure. But it is a plausible hypothesis, supported by all of the data that I am familiar with, that favorable energy supply conditions encouraged imaginative technological developments in the ways that energy was applied in production, which otherwise probably would not have occurred. Paradoxically, we may have been getting more from energy precisely *because* we had plenty of it and increasingly in the most desirable forms.

THE LATEST DATA

I would like now, briefly, to compare the long historical record, just reviewed, with what has been happening more recently. It is a well-known fact that the intensity of energy use in the United States has fallen sharply ever since the oil embargo year of 1973. Btus of energy consumed per constant (1972) dollar of GNP have fallen from 59,400 in 1973 to 53,300 in 1979, 51,300 in 1980, and an estimated 49,000 in 1981.[8] Between 1973 and 1979 this equals an average annual rate of decrease of about 1.8%, which, although substantial, is well below the annual rate of decline during, for example, the early postwar years of 1948–1953, when no one was paying any attention to energy productivity. However, following the second oil price shock that was triggered by the Iranian revolution, the average annual rate of decline in energy intensity accelerated to an unprecedented 4.1% between 1979 and 1981.

While energy productivity has been improving at a very high rate during the past decade, the overall productive efficiency side of the story has been highly unfavorable, and has become a matter of great concern. The post-1979 years that witnessed a new high in the rate of growth of national energy productivity, also saw a decline in productive efficiency with a *fall* in total factor productivity of about 0.3% per year between 1979 and 1981.[9]

In juxtaposing these two measures of performance since 1979,

I do not mean to say that they are causally related. Indeed, the evidence yielded by quantitative economic research assessing the linkage between deteriorating energy supply conditions and declines in the growth of productivity during the 1970s gives an ambiguous answer. Standard growth accounting analyses by economists assign only a minor role to energy in the productivity slowdown; some econometric results, on the other hand, show energy supply developments to be a crucial element in the explanation. The entire question concerning energy's role in the productivity slowdown is still under intensive study, but, as of now the answer is up in the air so far as the economics profession is concerned.

Nevertheless, it is clear that something has been happening to productive efficiency in recent years which is a matter of great concern, As Solomon Fabricant, one of the most careful students of long-term productivity trends in the United States has put it:[10]

> 'By either of the definitions of productivity in common use, with any of the measures that deserve attention, against all relevant historical standards of comparison—it is evident that recent productivity growth has been disappointingly slow. . . .'
>
> '. . . Increase in labor productivity has generally been not only an important source of the economic growth of the United States, but the *dominant* source. When labor productivity growth falls off and more worker-hours per capita become dominant—as is currently the situation—there is cause for worry.'

Although there is reason to be heartened by the sharp improvements in energy productivity, we cannot remain unconcerned about the dismal performance of broader measures of productive efficiency, especially at a time when solutions to many of our most difficult economic and social problems are seen to depend upon strong improvements in productivity and growth. The American economy is nowhere near matching its earlier performance during those years when productive efficiency and energy productivity were together growing substantially, years when energy abundance and flexibility appear to have been important facilitating factors.

CONCLUSIONS

Let me now state a major theme that emerges from the historical record. *Relative to the output of final goods and services*, energy consumption can be lowered by *reducing* the amount of energy consumed within particular processes of production (an important element in recent experience), and also by *expanding* the output of goods and services produced through raising the overall efficiency of productive operations (a significant factor in the earlier record). Except in the recent past, the special strength of energy as a production factor lay in its ability to support the growth in output relative to inputs of labor, capital, *and* energy through its synergistic relationship with technologies that led to improvements in overall productive efficiency.

I do not want to be misunderstood. I am not saying that innovative energy-using technologies were the sole cause of rapid improvements in overall productive efficiency. However, I do believe they were a major cause. Furthermore, I am not saying that energy supply developments were the sole cause of the emergence of innovative energy-using technologies. What I am saying is that energy supply developments were *essential* features of this process and that two particular supply developments were of critical importance. First, an abundant energy availability on favorable terms encouraged the development and spread of new technologies which favored the use of energy relative to labor and, to a lesser extent, the use of energy relative to capital. These new technologies were often able to raise the combined productivity of labor and capital. Second, the energy forms especially effective in achieving such productivity-enhancing effects during much of the 20th century were electricity and fluid fuels. The key mechanism involved appears to have been technical change which occurred in response to quality features of these energy forms. What was being sought in these technical changes were reductions in costs for total productive systems, not necessarily reductions in energy costs themselves; but, on balance, improvements in energy productivity, as well as in labor and capital productivity, turned out to be a significant result.

The long historical record in the United States shows that in the past conditions of energy supply were favorable to outcomes that permitted the growth in both energy productivity and overall productive efficiency. One cannot be certain whether or not future supply conditions will be conducive to such outcomes. My own judgment is that the richness of the remaining energy resource base in the United States (and probably world-wide, as well) is such that proper policies could yield favorable future supply outcomes even though comparative energy costs will be higher than in the past.[11]

However, even with such a favorable energy supply situation, one might legitimately ask whether the special relationship that once existed between the technologies of energy use and the conditions of energy supply can be replicated, or whether they were unique to an earlier era in which technical advance is viewed as having consisted mainly in energy and machinery replacing human muscle and animal power. Here, too, history offers some useful insights which may be illustrated with the example of industrial electrification.

Change in energy use technology, as exemplified in the electrification of manufacturing operations, did far more than replace human muscle. Instead, by supporting major improvements in already existing machine operations, electrification served as an important management tool for achieving a radical reorganization of industrial production technologies. Entire systems of production were altered.

Electrically-based techniques appear to have comparable potentials for the future in the increasingly important activities of the commercial, service, and information sectors of the economy, as well as in manufacturing. Such characteristics as susceptibility to precise control, highly focused application, fractional use, and a unique linkage to such technological systems as computers and robotics, strongly suggest that electrification may support radical changes in broad systems of production employed in these sectors.

REFERENCES

1. Sam H. Schurr and Bruce C. Netschert, with Vera E. Eliasberg, Joseph Lerner, Hans H. Landsberg, *Energy in the American Economy* (Baltimore: Johns Hopkins Press for Resources for the Future, Inc., 1960).
2. Harold J. Barnett, *Energy Uses and Supplies, 1939, 1947, 1965* (Washington: US Bureau of Mines Information Circular 7582, October 1950).
3. Data for the period of declining energy intensity show an annual average rate of decline of 1.3% in the energy/GNP ratio between 1920 and 1953. In contrast, between 1880 and 1920 the intensity of energy consumption (mineral fuels and hydropower) relative to GNP increased at an average annual rate of 2.2%.

   The exclusion of fuelwood from our basic energy consumption statistics barely affected the energy/GNP ratio during the period of declining intensity with which this paper is concerned. However, it had a major effect during the period of rising intensity; with fuelwood included the annual average rate of increase went up by a slight 0.3% per year compared to the 2.2% for mineral fuels and hydropower alone.

   There are solid grounds for believing that long-term energy/GNP comparisons are distorted less by excluding fuelwood than they would be by its inclusion. The reason turns on the overwhelming use of fuelwood within households as opposed to the productive sectors of the economy. Some 95% of all wood consumption in 1880 was used for household purposes, largely space heating; thus with fuelwood included, some two-thirds of the *total* fuel supply in 1880 was consumed in households. (See, *Energy and Economic Growth in the United States*, by Sam H. Schurr and Vera E. Eliasberg, [Washington: Resources for the Future, Inc., Reprint no. 35, August 1962]).

The energy total for mineral fuels and hydropower does not pose this problem. Estimates indicate that in 1880 about one-quarter of this total was used in households and the comparable percentage is not much different even today.

4. Schurr and Eliasberg, *op. cit.*, p. 11.
5. See, Warren D. Devine, Jr., *Historical Perspectives on the Value of Electricity in American Manufacturing* (Oak Ridge Associated Universities, Institute for Energy Analysis, draft Research Memorandum, May 1982).
6. In a subsequent article, 'Electrification and Capital Productivity: A Suggested Approach' (*The Review of Economics and Statistics* 48 (No. 4) 1966), Richard DuBoff presented further evidence supporting the linkage between the growth in electrification and the spurt in the growth of productivity in American manufacturing following World War I.
7. *Energy, Productivity, and Economic Growth*, Proceedings of a workshop held at the Electric Power Research Institute, January 12–14, 1981 (to be published in 1982).
8. US Department of Energy, *Monthly Energy Review*, p. 14, (April 1982).
9. *Energy, Productivity, and Economic Growth, op. cit.*
10. Solomon Fabricant, 'The Productivity-Growth Slowdown: A Review of the "Facts".' in *Energy, Productivity, and Economic Growth, op. cit.*
11. This judgment is based upon data and analyses presented in *Energy in America's Future*, by Sam H. Schurr, Joel Darmstadter, Harry Perry, William Ramsay, and Milton Russell (Baltimore: Johns Hopkins University Press for Resources for the Future, 1979). However, this same study also describes the numerous serious obstacles standing in the way of achieving a political consensus in support of policies that would be required for attaining favorable supply outcomes. (See, in particular, Chapter 1: "An Overview and Interpretation".)

PART IV

The Future for Oil

Abstracts (see Section 5)

Chapter 7

Establishing an Oil Price Window: The Influence of Policy on Future Oil Prices

Henry D. Jacoby and James L. Paddock**

Our task is to assess the opportunities for energy 'policy' in the face of the changing world market. To help set the stage we would like to explore the limits of policy intervention—and our knowledge of its effects—and make an argument about where our attention should be directed.

To talk about policy, of course, we need to be clear about objectives—not necessarily in agreement among one another, but clear. We will take the importer-nation's perspective, and propose that our objective with respect to international oil is to try to ensure that oil market conditions do no more damage to our economic health, and growth, than is necessary given the depletion of low-cost sources. This objective applies to developed and less-developed countries alike; and it implies a preference for lower oil prices as opposed to higher, limiting the transfer of wealth to exporters and lowering the drag on our economic performance. This objective implies a particular aversion to sharp jumps or spikes in the oil price, with their perverse effects in terms of inflation and economic contraction.

What evidence do we have on which to base expectations for the oil scene for the remainder of the 1980s, and beyond, and about

*Sloan School of Management and Energy Laboratory, Massachusetts Institute of Technology.

the role 'policy' can play? Clearly this is dangerous ground, for the field is littered with the remains of forecasts, and policy analyses based on forecasting exercises, which failed to capture the complex dynamics of this market. Indeed, the uncertainties are so great, and our ignorance of key relationships so deep, that we will avoid the now-common approach to analysis, which uses base-case or reference forecasts and the exploration of bands of error. Rather, we take the approach of ruling out certain combinations of future events that we can plausibly argue are 'unlikely', and then see what is left. The procedure defines a 'window' into the future which consists of the remaining set of non-unlikely outcomes, and this result is offered as a way of stating the existing level of knowledge, and ignorance, about likely future developments.

It is an approach, we feel, which gives a more true impression of the limits to our knowledge, than does analysis based on single-path scenarios of 'reference', 'high', and 'low' conditions. The issue of policy intervention then becomes a discussion of ways we might influence the dimensions of the window, and where the world ends up within it.

To illustrate the approach, and prepare for discussion of policy questions, we borrow from a set of calculations performed recently by members of the MIT World Oil Project (Jacoby and Paddock, 1981) which covers the period to 1990. The analysis involves the definition of a universe of possible combinations of oil price and world economic conditions over the decade, and then the use of analytical models and judgment to define the set of points, G, that can be shown to be unlikely to occur. Several factors come into play. The oil market must clear at each point along the way: consumption cannot exceed the quantity of oil made available to the world market. The effects of oil price on world economic growth must be properly accounted for: there are oil market conditions under which healthy economic growth probably cannot be sustained. And, of course, there is the behavior of the OPEC leaders themselves. What follows then is a set of calculations, some rather mechanical, which bring these various factors together to rule out the 'unlikely' combinations and identify the window or envelope of feasible paths to the future.

Once our vision of the window has been constructed, we can return to the questions of policy, and how interventions of various sorts may influence the outcome.

THE OIL MARKET WINDOW IN 1990

To carry out the analysis, we need to be able to analyze the responses of oil demand and supply to conditions in the oil market. For this purpose we make use of a set of studies, estimates, and model-construction activities carried out over several years by the MIT World Oil Project. The demand simulation model forecasts energy demand by sector in each of the OECD countries and aggregates their petroleum product demand with crude oil demands from the rest of the world. The equation structure has been estimated econometrically and the resulting elasticity and fuel share relationships form the core of the model. The supply model forecasts possible production capacity for all oil-producing countries with an on- and off-shore division for several. It is essentially an inertial-process type model which combines geologic and petroleum engineering characteristics within an economic analysis. The details of the simulation model are described in Carson, Christian and Ward (1981).* The econometric work behind the demand model has been documented by Pindyck (1979), and the supply analysis framework is detailed in a paper by Adelman and Paddock (1980).

The purpose of the exercise is to analyze oil prices and economic growth, and to look ahead to changes between now and 1990. Of course, where the world ends up depends strongly on the number and severity of the disruptions along the way. There are many circumstances that could lead to oil price shocks and extreme deviations in economic performance, and one cannot contemplate analyzing them in any detail. Therefore, for the sake of some numerical calculations we begin with a smoothly changing world. Our calculations assume a 1980 world oil price, P_t, (in mid-1980 dollars) of $30 per barrel, and consider alternative paths of exponential growth of this price. Similarly, we first treat the change in gross domestic product, GDP_t, as a smooth exponential process over the period of analysis. Later we will look at the effect of oil price shocks.

Oil Prices and Market Clearing

In defining the price/GDP growth window in 1990, the first fact we know is that oil markets must clear all along the way; it is not possible for demand to exceed capacity. Moreover, we define 'market clearing' as a condition wherein demand and supply are equilibrated at an assumed oil price *without* creating conditions where prices

*The research behind these models has been supported by the US National Science Foundation under Grant No. DAR 78-19044. However, any opinions, findings, conclusions or recommendations expressed herein are those of the authors and do not necessarily reflect the views of NSF. The work also has been supported by the MIT Center for Energy Policy Research.

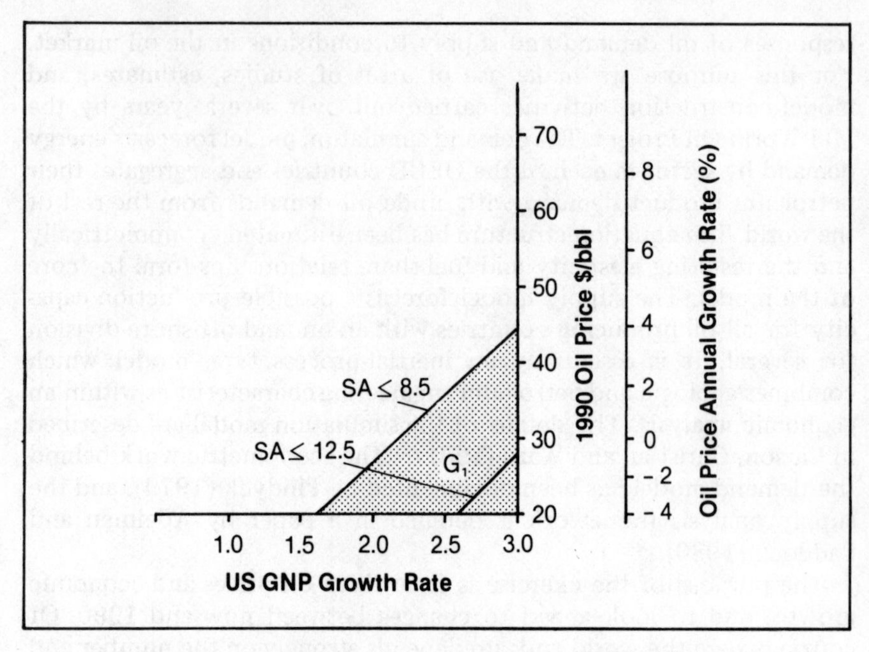

Figure 7.1. Oil price and GNP growth (1980 dollars).

might explode due to small perturbations in the market. Figures for 'capacity' are approximate and may vary with seasons and the condition of reservoirs and equipment; similarly, demand may vary due to seasonal factors, weather and stock-holding behavior. A stable market-clearing price really implies some slack in the system to handle these departures from forecasts of annual capacity and demand.

Therefore the definition of market clearing includes an excess of capacity over demand of 2 mbd at all times. To separate feasible from infeasible combinations of economic growth and oil price, the simulation model is run for combinations of growth and price which just barely satisfy the clearing constraint. By a sequence of solutions of this form, we can identify a subset of those pairs of vectors—each rising at an exponential rate—which constitute the boundary of the constrained set.

The resulting boundary defines the lower sill on the price-growth window, as shown in Fig. 7.1. The vertical axis shows the oil price (in constant 1980 dollars) that will hold in 1990; the range spans from a little less than $30, the 1980 level, to over two times that level. Also shown is the implied constant annual percentage growth rate in real prices, taking $30 in 1980 as the base, which yields the 1990 price shown (in 1980 dollars).

The horizontal axis shows varying levels of GDP growth, with the

US used as an example. At the rightmost extreme (at the intersection with the vertical axis) is the 'base' growth rate. In these calculations, the 'base' rate for the US is 3.0 percent growth; for Japan it is 5.0 percent, France is 3.0 percent, the UK is 2.5 percent, the FRG is 4.0 percent, and other Western Europe is at 3.2 percent. The scale moving left indicates (lower) growth rates that may actually be realized over the decade. The growth of all countries is assumed to move up and down in proportion, and so the US values in Fig. 7.1 are also a representation of what is assumed to be happening to the growth rates of all countries.

The line labeled '$SA \leqslant 8.5$' represents the boundary (the lower sill of the window) of the feasible pairs of (P, GDP) under the assumption that Saudi Arabia will not produce above 8.5 mbd. Thus the shaded area in Fig. 7.1 indicates a set of (P, GDP) pairs that we judge 'unlikely'. Points in that shaded area are not impossible, for our demand and supply models may be in error. But to our view, these points would be unlikely to be realized in markets that clear to 1990, given the Saudi constraint. The area to the upper-left of the line $SA \leqslant 8.5$ is then the set of price-growth circumstances that cannot be judged unlikely simply because of a need to clear the oil market. Since we are using the language of 'sets' to define the pairs of vectors (P, GDP), the shaded area, now ruled unlikely with an 8.5 mbd Saudi policy, is denoted as the set G_1. Ultimately, the price/growth window will be what is left when all the sets of unlikely events have been ruled out.

Figure 7.1 also shows another line which locates the boundary of G_1 should the Saudi production limit be increased to 12.5 mbd. This line is closer to the origin since, with more oil available, higher rates of economic activity and lower price increases are feasible. In the discussion below, the possibility of arranging such a shift (or other market changes that have the same effect) will be an important policy option.

Oil Prices and Feasible Economic Growth

A second set of limits on likely combinations of (P, GDP) is imposed by the fact that GDP itself is influenced by the oil price. In terms of our analysis, there is a set $G_2 = f(P, GDP[P])$ which is unlikely for reasons of the depressing effects of energy price increases on economic growth. We have as yet no formal model that will support a calculation of the boundaries of G_2. Therefore we construct an exercise which will allow the use of as much formal analysis as exists, and permit the easy introduction of judgment about these relationships.

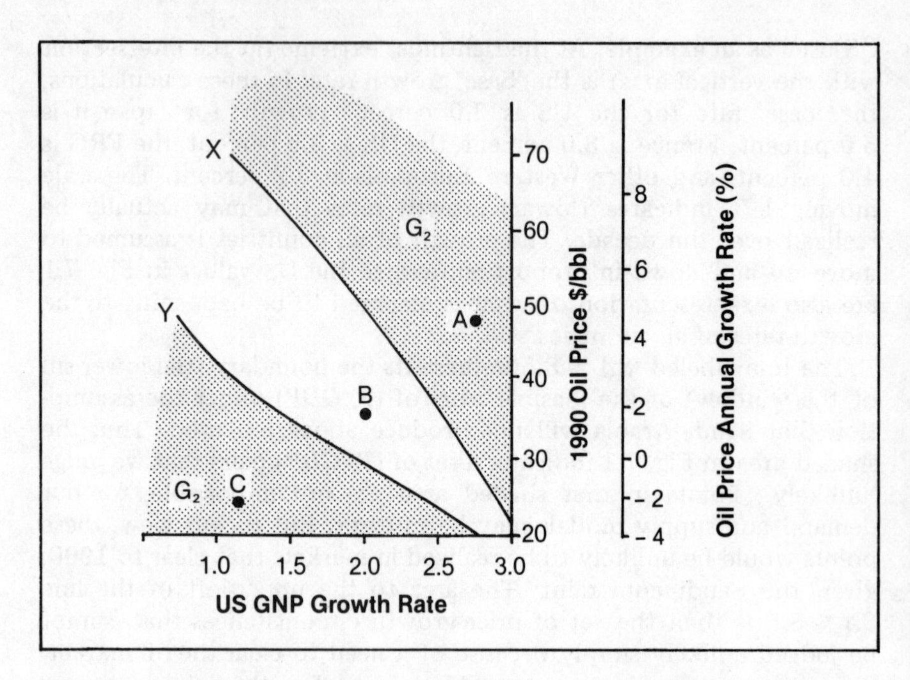

Figure 7.2. Oil price and GNP growth (1980 dollars).

The approach is illustrated in Fig. 7.2 which consists of the same price-growth space as Fig. 7.1. Take point A (in Fig. 7.2) for example, and ask the following question, 'Is it likely that GDP growth can be sustained at 90 percent of the base levels if oil prices are rising at 7 percent per year?' If the answer to that question is 'no', or 'not very likely', then point A falls into the region G_2. The same question could be asked about point B. 'Is 60 percent of base growth rates likely while prices are rising at 4 percent per year, to $45 per barrel in 1990?' Here the response may be, 'I see no reason why not'. Or, in our terms, point B is 'not unlikely'. By a sequence of such questions—with reponses perhaps aided by formal analysis of the world growth process and price elasticities of income—the upper line, labeled X in Fig. 7.2, can be located, and the upper region G_2 of unlikely combinations defined.

Thus, Fig. 7.2 shows our judgment of the upper end of expected economic performance. Our impression is that the base growth rates are unlikely to be realized, even if oil prices stay at 1980 levels. Furthermore, if real oil prices were to increase at 8 to 10 percent per year over the decade, it is unlikely that world economies could sustain growth greater than about 60 percent of base levels. The lower boundary of this unlikely upper portion of G_2, denoted by line X, reflects the limits on how *well* the world is likely to be able to do between now and 1990.

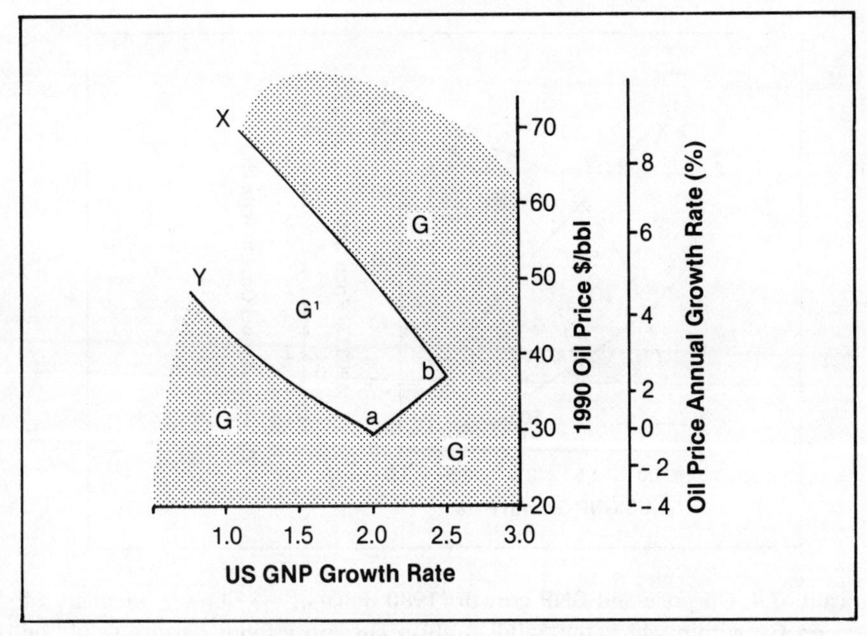

Figure 7.3. Oil price and GNP growth (1980 dollars).

Of course, there are some limits to how *bad* growth performance is likely to be, even given rising oil prices. Thus there is another portion of the set G_2 which represents downside prospects that are also judged 'unlikely'. This portion of G_2 may be described by the same process as before, only now comparing points like B and C in Fig. 7.2. Point C falls in G_2, meaning that we judge it unlikely that realized growth over the decade will be as low as 40 percent of base levels, if oil prices rise only by 2 percent per year on average over the period. Line *Y* depicts the limits on how badly the world is likely to do.

In Fig. 7.3, G_1 and G_2 are superimposed to present the union (shaded) of the two sets of conditions, $G = G_1 \cup G_2$, which the analysis shows is unlikely to come about over the next decade. If one expects the members of the OPEC core to produce near capacity and Saudi Arabia to produce up to 8.5 mbd whenever demand conditions call for it, then the likely combination (P, $GDP[P]$) is somewhere in the neighborhood of the line a–b in Fig. 7.3. This neighborhood is both feasible and consistent with the analytic implications of our model.

Unfortunately, that is not the end of the story. In our set notation, we have defined a set *G* which is unlikely; the complement G'—which contains the combinations ruled *not*-unlikely—is unbounded to the upper-left as shown in Fig. 7.3. That is because our supply model predicts the *capacity* of the core countries, not their actual production; and the line a–b in Fig. 7.3 is for a Saudi maximum of 8.5 mbd—

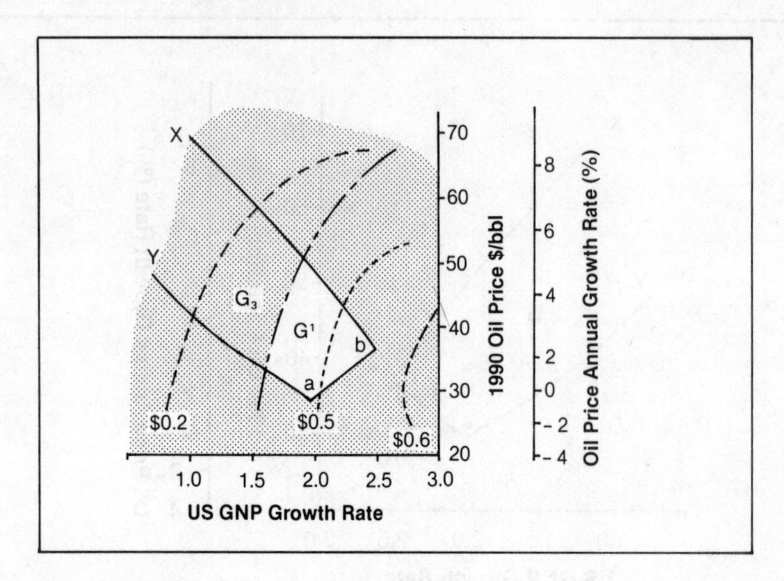

Figure 7.4. Oil price and GNP growth (1980 dollars). · · · · Lower boundary of region G_3 within which the Saudi Arabian current account balance would be negative in some years. — — — Loci of equal present value of additions to Saudi financial balances through 1990 ($ trillions).

a number which they can lower if they choose. Or other OPEC core countries can hold excess capacity. The feasible 'window', beginning at line a–b, has no top on it. To seek an upper bound one must turn to analysis of the behavior of the nations of the OPEC core.

OPEC Core Country Behavior

There are a number of hypotheses about OPEC core behavior. They vary according to one's view of the relations among the members of the core, and the objectives that these countries are likely to pursue over the decade. One minimal hypothesis is that over the decade Saudi Arabia will try to avoid going into deficit in the balance of trade in each year. That is, the Saudis will not hold great excess capacity (thereby holding up prices) if their foreign exchange revenues are not sufficient to cover import bills. The simulation model can be used to calculate that range of conditions (P, GDP) which, combined with a forecast of Saudi imports, will yield a zero balance on current account in the most stringent year. The results are shown in Fig. 7.4. To the upper-left of the heavy dotted line, Saudi net foreign financial balances must decline in some years in order to finance an excess of imports over oil revenues. This calculation thereby defines a set G_3

which is unlikely (not impossible, but unlikely) to occur. The set G of unlikely events can then be defined as the union: $G = G_1 \cup G_2 \cup G_3$, and the complement G' defines a bounded envelope or 'window' of likely oil price and world GDP growth combinations over the decade.

At this point, one of the most important conclusions of the analysis can be drawn: the window of prices that are 'not unlikely' in 1990 is very large. By our analysis it stretches from as little as zero percent annual growth from a $30 1980 price, to as high as 6 percent per annum over the period, or something in the neighborhood of $55 per barrel (in 1980 prices) by 1990. Prices outside this range seem unlikely; but *within* the likely set G' there is only a much weaker set of arguments as to where the outcome will ultimately be. Narrow-band forecasts of price to 1990 are not to be believed.

In this analysis we have divided events into only two categories, unlikely and not-unlikely. One can imagine a more complex definition of G', which defines zones of relative likelihood, perhaps by use of probability distributions of demand elasticities and drilling rates. We will not attempt that here. However, in the case of Saudi behavior, there is one extension of the financial balance hypothesis which is useful and easy to do. The dashed lines on Fig. 7.4 show the loci of equal additions to the present value (iso-present value lines) of Saudi foreign financial balances over the decade to 1990. The balances are discounted at a 2 percent real rate and stated in trillions of 1980 dollars. The closer the overall outcome to the lower right-hand corner of G' (at point b), the higher the Saudi addition to financial wealth during the 1980s.

That is, if the Saudis' objective were to try to maximize the present value of additions to financial wealth over the decade, they would set prices on a trajectory to around $37 in 1990 (2 percent growth) and hope the world will end up at point b over the decade, with GDP growth at about 80 to 85 percent of base rates. There is a continuing debate over Saudi motives, but to the extent one believes the hypothesis of wealth maximization, a distribution of likelihood can be imposed on G', and it will be a positive function of the wealth lines shown in Fig. 7.4.

THE EFFECT OF OIL PRICE SHOCK

Thus far we have built the analysis on the assumption that oil price increases come in a smooth exponential form, such as the pattern that might hold if OPEC were to follow the long-discussed price formula. Of course, experience is to the contrary. The events of 1973-74 and 1979-80 produced sharp price shocks followed by

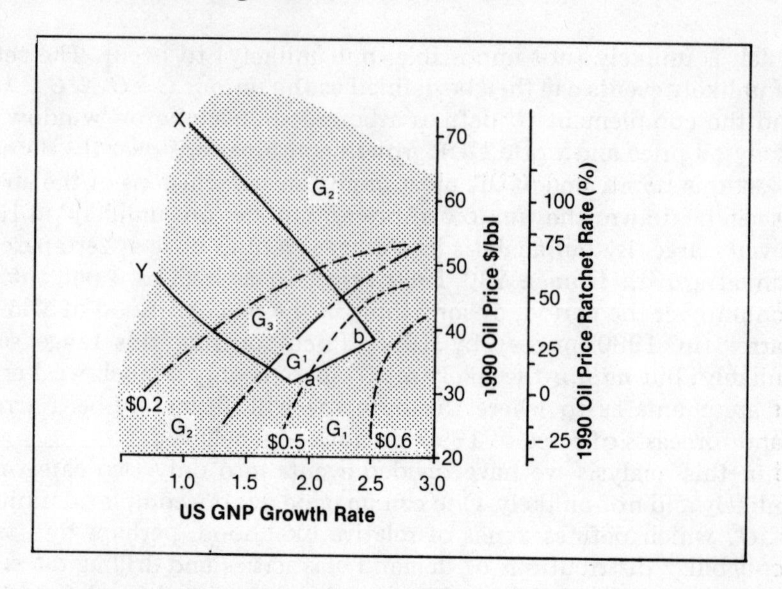

Figure 7.5. Oil price and GNP growth (1980 dollars). · · · · Lower boundary of region G_3 within which the Saudi Arabian current account balance would be negative in some years. — — — Loci of equal present value of additions to Saudi financial balances through 1990 ($ trillions).

periods of relative price stability in real terms as the higher price level was absorbed. Very likely the future holds further events like these.

Our analysis cannot fully explore this phenomenon, but it can shed some light by looking at the effect of a shock on the price window in 1990. Between 1980 and 1983 the world oil price is assumed to hold constant (in real 1980 dollars) at $30 per barrel. World economic growth is assumed to move from recession to recovery in 1982. The price shock comes in 1984, and world GDP growth rates (again presumed constant) over the period 1984 to 1990 are different as a result. The calculation procedure described above for the smooth process is followed, and G_1 is calculated on the assumption that real oil prices hold constant through 1990 at their post-shock 1984 level.

The results are shown in Fig. 7.5. Again the vertical axis shows the price in 1990 (which is in effect from 1984 to 1990), and the horizontal axis shows US economic growth performance for 1984 to 1990, after the shock. As before, all country economic growth rates change proportionately. We have drawn in our impression of the limits to post-shock growth economic performance, noted as G_2 again, and the heavy dotted line representing the Saudi current account balance constraint is also shown, identifying G_3.

One may argue about the $GDP[P]$ relationship, and Saudi behavior in the face of balance-of-payments deficits, but there is no doubt that the introduction of a shock makes a significant difference from the former analysis under smoothly changing prices. The upper limit of 'non-unlikely' prices in 1990 is reduced from the case with smooth growth in price and GDP. The analysis shows that a 1984 price ratchet of up to nearly 50 percent (say, to about $45 in 1980 dollars) can be sustained to 1990, but a ratchet greater than this likely would not. As before it also is not-unlikely that markets could clear to 1990 with no ratchet at all. That is, the bottom of the window is near the current price level.

Measures Affecting the Boundaries of the Window

With our way of viewing the future, as summarized in Figs 7.4 and 7.5, the role of policy falls into two obvious categories. First, there are those measures that influence the dimensions of the window itself. For example, increasing Saudi or other OPEC core capacity, relaxing a Saudi production constraint, or increasing capacity in non-OPEC areas, will lower the line a–b in Figs 7.4 and 7.5, making lower prices feasible in terms of market clearing. A similar effect could be gained by interventions on the demand side, which might (for a given level of economic activity) lower demand below that computed by our economic analysis. A glance back to Fig. 7.1, where a 40 mbd difference in Saudi electricity is shown, will give an idea of the magnitude of the effect.

Note that our analysis would not support an argument that such measures will actually lead oil prices to *be* lower, only that there is opened up a lower range of price one cannot rule out as 'unlikely' based on the requirement of market clearing.

Policy attention may also be directed to insulating economic activity from the effects of oil price increases—say, by means of more effective coordination of fiscal, monetary and trade policies. But note that success in this realm brings both good news and bad. Such measures will shift the line b–X to the right in Figs 7.4 and 7.5, opening up the window, and exposing an even higher region of oil prices that one cannot rule out as 'unlikely'. In effect, the more resilient the economies of the big importers, the higher the oil price the exporters can extract.

Attempts may also be made to affect the spending behavior (and thus the current account balances) of Saudi Arabia and the other surplus countries. If their appetites for oil revenues (or financial wealth-holding) were dulled, then the heavy dotted lines in Figs 7.4 and 7.5 would shift outward, and a higher region of sustainable prices would be opened up. If consumers could take actions that kept spending needs high, then by this analysis some higher-price regions could be ruled unlikely. Here is another double-edged sword, of course, for in some countries at least, high expenditure levels may well be de-stabilizing, particularly if arms purchases are a large part of the flow.

This review of the options leads us to the view that it is very difficult to reshape the window by policy means at our disposal. This is not to argue that attempts to shift the boundaries are not worth trying—particularly measures that lower the line a–b by encouraging capacity creation and production (inside OPEC or out) or facilitate adjustment of oil demand to higher oil prices. But the ability of consumer countries to influence the supply decisions of major exporters seems very limited, and the effects of government 'conservationist' intervention also appear small, in relation to the effects wrought by the price itself. And unfortunately, even if we were able to achieve, by 'policy', a lowering of the bottom of the window, it is hard to make a convincing argument that there would be any very direct influence on the oil price we will likely see in 1990.

Measures Affecting Where We Come Out in the Window

If, as we argue, there are limits to our ability to influence the window, how about ways to make the favorable regions more likely than the unfavorable ones? On this point, the discussion varies depending on one's view of the nature of price setting in the market. If one expects the smooth, controlled evolution of prices, as illustrated by Fig. 7.4, then the issue is our ability to influence OPEC leaders as they might try to manage that process. Our analysis suggests that *given a level of capacity creation and willingness to produce*, our interest and the Saudi interest (or, at least, our version of the Saudi interest) are the same. That is, to go through the window near point b. But once again it is not at all obvious that we share a common view of one another's interests, or that our ability to convince and/or influence the outcome can be counted on for much.

Of course, the question of influence over exponential price trajectories is probably moot, for the more likely mechanism of price change is not by smooth managed adjustment but by poorly controlled price shock. To analyze that case let us go back to Fig. 7.5. It shows the level of price shock, if it occurred in 1984, that could be sustained to 1990. (Bigger shocks are possible, of course, but it is likely a portion of the price increase would be eroded in real terms, as is happening today.)

It is a situation which lends an essential randomness to the outcome. In the case of smooth price adjustment, one could at least imagine a rational price-setting process—although we argue we can't forecast it. But in the shock case even that fiction is unsustainable. Where the world comes out in the window is very much a function of random events—disruptions in oil supply mainly, which trigger inventory panic and price escalation. With the phenomenon of the 'ratchet' (Jacoby and Paddock, 1980) the prices once bid up do not come back down, or only do so very slowly.

Here we have an opportunity for policy intervention to make a difference. If we can moderate the influence of these random shocks, we would stand a better chance of avoiding the low-growth regions of the Fig. 7.5 window. We should not be naive, of course. There is evidence of exporter manipulation of these shock situations, even if they did not create them; and the influence of price stabilization measures can be overwhelmed by pro-shock supplier responses. Still, it is the area of government action that holds the most promise for improving conditions over the rest of the decade.

For this purpose, the most important instrument appears to be the policy regarding stockpiles—not just their existence but the procedures for their use. To date there appear to be almost no plans or procedures for how oil stocks (those under 'policy' control, anyway) are to be used in an emergency. If anything there is a general notion that such stocks should be held back in small disruptions—in case something really terrible happens. There appears to be little government or industry support for the use of strategic stocks for oil price stabilization in disrupted markets.

Thus from our study of what we know about future oil market conditions, and what we might do about them by policy intervention, the stabilization question emerges as a most important topic for analysis, planning, and international agreement. We may not know which portion of the window we will pass through in 1990, but we do know the mechanics of the processes that drive us in the direction of its more painful regions.

REFERENCES

Adelman, M. A. and Paddock, J. L., *An Aggregate Model of Petroleum Production Capacity and Supply Forecasting*, MIT Energy Laboratory Working Paper No. MIT-EL 79-005WP, revised July 1980.

Carson, J. W., Christian, W. and Ward, G., *The MIT World Oil Model: Documentation and User's Guide*, MIT Energy Laboratory Working Paper No. MIT-EL 80-026WP, revised December 1981.

Jacoby, H. D. and Paddock, J. L., *World Oil Prices and Economic Growth in the 1980s*, MIT Energy Laboratory Working Paper No. MIT-EL 81-060WP, December 1981.

Jacoby, H. D. and Paddock, J. L., 'Supply Instability and Oil Market Behavior', *Energy Systems and Policy* 3 (No. 4), (1980).
Pindyck, R. S., *The Structure of World Energy Demand*, MIT Press, Cambridge, 1979.

PART V

Problems in Oil Refining and the Pace of Substitution

Abstracts (see Section 5)

Chapter 8

The Changing Economics of Oil Refining

*J. H. Culhane**

There have been drastic structural changes in the world economy, and they will have a dramatic effect on the economics of oil industry operations. We have to consider energy as a total system, and then examine what this means for the demand for oil in general, for different refined products in particular, what this means for the structure of the refining industry, and its likely effect on synfuels and exploration.

THE NEW ECONOMIC CLIMATE

I don't suppose that it will be any surprise to anyone that the demand for oil is likely to decline over the rest of the 1980s. We see this happening for three main reasons, which we have tried to model explicitly, but which are qualitatively self evident.

Firstly, the rate of economic growth in both the developed countries, which are the large consumers, and in the LDCs, is going to be less than it was during the 30 years after the war. In both respects we are more pessimistic than most of the forecasts I have seen. There is, in my view, a terrific segmentation of the world economy, which is perhaps most marked in the United States. In short, I expect significantly lower growth than we have become accustomed to in almost all sectors of the world economy.

*Vice President for Planning, Occidental Petroleum Corporation.

Secondly, the effects of two price shocks, first in 1974 and again in 1979, have triggered a steady increase in the thermal efficiency of converting energy to economic activity. This is not simply conservation, in the sense of eliminating waste, or the use of smaller cars, the more efficient use of energy in myriad facets of the economy, which has occurred as a response to higher prices.

Thirdly, also triggered by higher prices, is increased interfuel substitution generating power by nuclear energy and by coal, heating homes by gas rather than gas oil, driving cars on diesel rather than gasoline.

This really is a new climate; while the demand appears to be level—a far cry from the 7% annual growth which was once thought to be absolute truth—production outside the OPEC countries seems likely to increase somewhat.

With a steady level of production in the US, increasing production from Mexico, and modest increase from elsewhere in the world, the result for OPEC in general and Saudi Arabia in particular is quite sharp. We have therefore to expect declining liftings from the OPEC countries, which by 1990 seem likely to reach record low levels.

A simplified, but popular, model of OPEC, is the price leadership model with Saudi Arabia as the price setter and producer of last resort. For this model to be valid, true leadership will indeed be called for.

We have not tried to model the pricing policy for Saudi Arabian Light. We simply do not know enough about the relationship between price and long-term demand to make intelligent deductions, but we do know, which we didn't know a while ago, that the people who set the price do realize that a change in the price of oil does indeed change the pattern of demand. Over time, but how much is simply a guess, and our estimates may well be quite different from yours. I know of no satisfactory model of price setting, although both Gately and Singer have given us insights to the economic inputs which the price setters may receive.

This is the new economic climate. It marks a significant change in the price history of the last 7 years. It's not merely a 7 year itch, it's a 7 year scratch, and since it's a fairly deep scratch, there's going to be a certain amount of blood left on the floor.

PROBLEMS AND OPPORTUNITIES IN REFINING

Although oil demand is inexorably going South, the different parts of the barrel behave very differently and this will create major problems and opportunities for the refining industry. Gasoline, of course, will continue to go down somewhat in the United States, and to be basically level or maybe even rising a little, in most Europe. Gas oil seems to be exhibiting the reverse tendency, to rise in the United States and to decline somewhat in Europe. One thing common to both hemispheres is that resid is going out of the window, so that we can see the greatest change taking place in the structure of the refining industry.

In 1979, the peak year for demand, the utilization of distillation capacity was about 85% and cat crackers were running at 95% capacity. By 1981, both distillation and cat cracking were running at about 75% utilization. Hydro-conversion, particularly coking, was fully utilized.

By 1985, distillation will be down to 60%–65% and cat cracking down to about 50%. Given this situation, it seems to me extremely unlikely that the United States Government will place any restrictions whatsoever either on the export of gasoline or on the export of naptha as a petrochemical feedstock.

There should be a slightly different situation in Europe. In 1981, distillation in Europe was operating at around 53%, cat crackers were nearly fully utilized, and hydro-converters and cokers were working flat out. The results translate neatly into refining margins. Hydro-skimming was losing money, those fortunate enough to have a coker are laughing all the way to the bank.

We see an even greater change by 1990. Even allowing for all the new building, there is going to be a squeeze on gas oil and, particularly, on kerosene. European cat crackers are going to be fully utilized, and a cat cracker is absolutely necessary for the survival of a refinery.

However, the price of gasoline in Europe will be directly related to that in the United States, where there is a great deal of idle cat cracking, so that the margin will not be very great. I have no doubt that refineries which already have cat crackers will survive, but it is unlikely that hydroskimmers which have not got a cat cracker will be able to justify an investment of $300MM, and unlikely that they can survive the next decade without one, unless, of course, they have some geographical protection. This is not exactly good news, and I will no doubt pay for it when I go to the great hydrocracker in the sky. It's relatively easy to be pessimistic about refining when one has gotten out of the business.

The key variable in all this is economic growth. If you expect economic growth to be lower than it used to be, and interest rates to be higher, you can see that is why the synfuel industry has been set back a decade.

THE IMPACT ON EXPLORATION

More importantly, however, is the effect on exploration. As you know, most of the major oil companies are slashing their exploration budgets, and the reason is twofold. One is the falling cash flow from existing operations, when compared to expectations; the other is the declining value of small fields in remote locations.

Exploration costs, for the industry, are now something like $12–15 in the US and $6–8 per bbl in the rest of the available world; i.e. 40% or 20% of the value of the barrel.

In the United States, where the infrastructure of pipelines is in place, exploration is still profitable, particularly for companies with a large inventory of romantic acreage. Enhanced oil recovery is also still attractive. But for other parts of the world, where profits, if there are any, are split between the company and the host government on a somewhat different basis than the federal income tax, the size of the target field becomes critical, and the changed outlook for prices has drastically increased the minimum size of field that would be attractive enough to justify exploration.

PART VI

Comparative Energy Licensing, Leasing and Taxation

Abstracts (see Section 5)

Chapter 9

Efficiency in Leasing

Walter J. Mead†, *Asbjorn Moseidjord*† *and Philip E. Sorensen* ‡

Economic theory suggests that the goal of national policies for developing natural resources on public lands should be to maximize the in situ *present value of the resources. This goal puts economic efficiency at the top of the list of considerations which are typically involved in setting national policies for the development of oil and gas resources. The argument of economic theory is that if the* in situ *value of oil and gas resources (usually called economic rent) is maximized, the owners of these resources (the public) will enjoy the greatest possible improvement in their economic welfare.*

The system of bonus bidding with a fixed royalty requirement, widely used in the United States for leasing offshore oil and gas resources, is rejected by most of the governments of the world in favor of more direct government involvement. It is often argued that the number of bidders for offshore oil and gas leases is too small to yield competitive outcomes or, more generally, that the oil industry is too oligopsonistic to behave in a competitive manner at lease

*This research was supported by US Geological Survey Contract No. 14-08-0001-18678. The views expressed are those of the authors. The very detailed explanation of the methodology is given in the full paper available in the BIEE and IAEE Archives.

†University of California, Santa Barbara.

‡Florida State University, Tallahassee.

auctions. However, the US experience indicates that the number of participants in bidding for offshore leases is very large, amounting to date to more than 130 separate firms (see Wilcox, 1975). Furthermore, it would appear that the oil and gas lease market approximates the theoretical requirements of a 'perfectly contestable market' (Baumol, 1982). In the absence of entry barriers, competitive results may be achieved even under conditions of few bidders.

But the question need not be argued in theoretical terms alone. We have analyzed 1,223 OCS oil and gas leases in the Gulf of Mexico issued by the US Government over the years from 1954 through 1969. Our findings strongly support the conclusion that the offshore lease market is competitive and that rates of return to the investors in oil and gas leases are normal or subnormal. These oil and gas leases were issued using a cash bonus bidding system with a fixed $16\frac{2}{3}$ percent royalty requirement. The combination of strong incentives for efficiency in the development of these resources, low administrative costs, and normal or subnormal rates of return to the lessee firms, leads us to conclude that this system has effectively transferred to the government most of the economic rent implicit in these resources.*

In the sections which follow, we shall present a discussion of economic rent maximization (Section 1), alternative leasing systems from an efficiency perspective (Section 2), and our empirical findings regarding competition, and thus rent collection, under the cash bonus bidding system (Section 3). Some observations on policy issues will be made in a concluding section.

MAXIMIZING AND COLLECTING ECONOMIC RENT

In professional journals, one finds reference to *the* economic rent as if it were independent of the leasing method employed. However, we point out three classes of events associated with leasing which either reduce the available economic rent, or lead to dissipation of some of the rent which is collected. (1) If the leasing method imposes unnecessary costs on the lessee, then the amount of economic rent available for collection will be reduced. (2) If the leasing method imposes disincentives which reduce the level of production below the social optimum, the available rent will be reduced further. (3) If the method requires unnecessary policing or administrative costs

*Because the US system included a $16\frac{2}{3}$ percent royalty requirement, some economic inefficiency is present causing a reduction in the available economic rent.

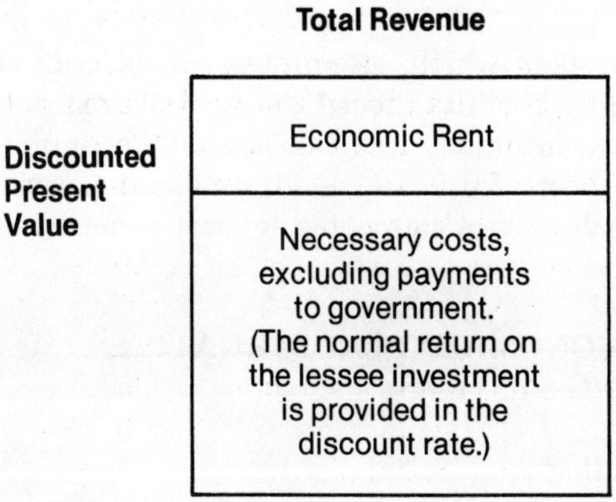

Figure 9.1. Model of Economic Rent Estimation

on the part of the lessor, then some of the economic rent that may be collected will be sacrificed and will not flow to the benefit of the resource owner.

In Fig. 9.1, we present a graphic model which identifies the economic rent component and clarifies the important efficiency issue. The vertical distance in the diagram represents the discounted present value of the total revenue obtainable from a given oil or gas lease when production from that lease is carried to the point where the social value of the incremental unit of production equals its incremental social cost. The discount rate represents the competitive return (or the risk adjusted opportunity cost) for investments in the lease. If competition for leases is effective, lessees in the aggregate will earn a normal return on their investments.

The largest segment of Fig. 9.1 represents the lessee's 'necessary costs' of exploring for and producing oil and gas from a property and does not include payments to the lessor or administrative costs incurred by the lessor. With all lessee costs and income flows expressed in terms of present values, the residual is the economic rent available for collection by the lessor. This rent may be collected in a variety of forms including a lump sum bonus payment, royalty payments, rent payments, or a profit share payment.

Two important efficiency issues related to alternative leasing systems can now be illustrated. First, any leasing procedure that leads to premature (delayed) abandonment of a lease implies that costs increase less (more) rapidly than revenue at the shut-down point. Thus, the economic rent segment in Fig. 9.1 is reduced when leasing method incentives lead to suboptimal decision making by the lessees.

Second, a system which discourages exploration or production investments whose benefits exceed their costs, or stimulates uneconomic investments, again reduces the available economic rent. This is particularly relevant for lease conditions and government-imposed regulations which are not supported by cost-benefit analysis.

THE ECONOMICS OF ALTERNATIVE LEASING SYSTEMS

Pure bonus bidding

Under this system, the lease is awarded to the highest bidder who makes payment in the form of a single lump sum at the time of the auction. In terms of allocative efficiency, this system has three major advantages. First, no part of the economic rent transfer is charged against a unit of production. Therefore, managerial decisions concerning whether to produce from a known discovery, whether to make an investment in enhanced recovery, and when to terminate production may all be made on the basis of real marginal costs and revenues and are not burdened by transfer payments that affect marginal costs. Second, the most cost-effective firm is likely to be the highest bidder. And third, lease administration costs are minimized and therefore economic rents are maximized.

Critics of the cash bonus bidding system generally do not question its allocative efficiency, but rather its ability to actually transfer the full amount of economic rent to the lessor government. Their arguments are usually in terms of inadequate competition at lease auctions. In Section 3, we review these arguments and state our empirical findings which, in essence, are that the cash bonus bidding system *is* an effective means for capturing economic rent.

Pure royalty bidding

The two major advantages of royalty bidding are, first, there is a one-to-one relationship between production and payments to the lessor, and second, no front end payments are required.

Fig. 9.2 illustrates the basic problems of royalty payments as a method of transferring economic rent. As a reservoir is produced, real social (and private) costs per barrel increase due to declining reservoir pressure and declining reserves. From an allocative efficiency perspective, production should be continued until incremental costs rise to

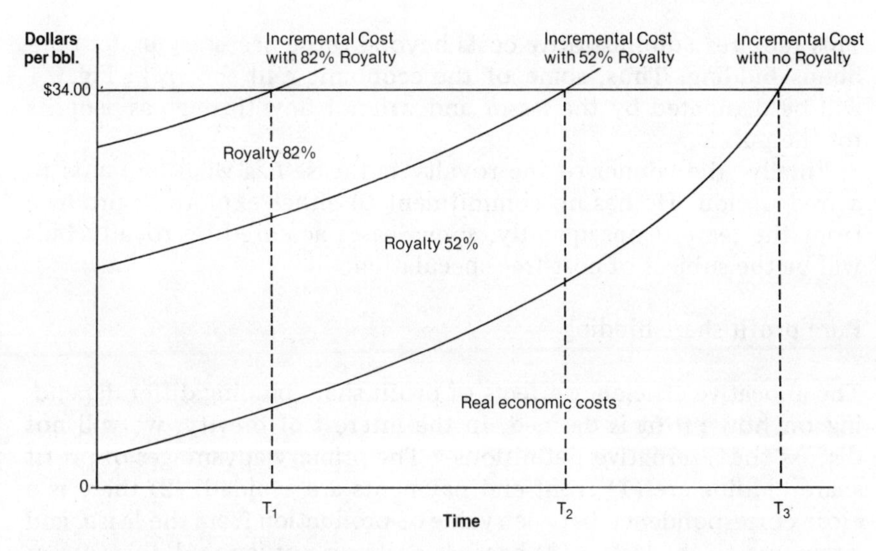

Figure 9.2. Model of Oil or Gas Well Abandonment Points Under Various Royalty Rates.

Figure 9.2. Model of oil or gas well abandonment points under various royalty rates.

equality with incremental revenue. This optimal closedown point is shown at T1. Where a royalty payment is used as a means of transferring economic rent, this transfer payment (not a real social cost) becomes an addition to the incremental costs to the operator and leads to premature abandonment of productive wells. The record of royalty bidding for OCS leases in the United States shows bids ranging from 52 to 82 percent of wellhead value. Given crude oil prices at about $34/barrel, Fig. 9.2 shows that leases would be prematurely abandoned at T2 and T3.

Also, it is obvious from Fig. 9.2 that, given the fact that royalty payments are assessed against gross income, some otherwise economically productive wells will be plugged and abandoned prior to production because the net incremental revenue after all costs, including royalties, fails to cover development costs. Further, some investments in enhanced recovery, which would be profitable on the basis of social costs and benefits, will become unprofitable when they must bear the added burden of royalty payments. The three allocative efficiency problems discussed above reduce the present value of total revenue shown in Fig. 9.1 and therefore sacrifice some of the economic rent available to the public. Although it is a minor point, where the basis for rent collection is a royalty payment, the lessor must monitor production to ensure collection of royalties.

This requires administrative costs beyond those necessary under pure bonus bidding. Thus, some of the economic rent shown in Fig. 9.1 will be dissipated by the lessor and will not flow through as benefits for the public.

Finally, the winner of the royalty bid lease has what amounts to a free option. He has no commitment to either explore or produce from the lease. Consequently, some leases acquired by royalty bids will be the subject of cost-free speculation.

Pure profit share bidding

The allocative efficiency effects of profit share bidding differ depending on how profit is defined. In the interest of brevity, we will not discuss the alternative definitions.* The primary advantages of profit share bidding are (1) front end payments are avoided, (2) there is a close correspondence between value of production from the lease, and payments to the lessor, (3) because costs are not ignored, this system is preferred to royalty payments, and (4) it is conceivable that profit sharing may constrain over-zealous regulators and environmentalists from imposing costs that cannot be justified on the basis of benefit/cost analysis.

The major disadvantage of profit share bidding and payments is due to the heavy cost of administering the system. Administrative costs paid by the operator depress the economic rent. Additional administrative costs borne by the government absorb some of the rents that would otherwise be available to the public.

Excessive lease administration costs arise out of the following situations: (1) Where the lessee is an integrated oil company selling its production to itself, the sale is not at 'arm's length'. Consequently, the lessor must carefully police the 'selling price' to ensure that the public is getting the benefit of true market value. (2) Accounting procedures must be carefully policed to ensure that lessee company overhead costs are not unreasonably charged against lease revenue. (3) Where some of the benefits of investments in lease development spill over to the lessee company, the lessor must evaluate investment made by the lessee. (4) 'Gold-plating' (poor cost control) is likely to occur on profit share leases where the share paid to the government is very high and the retained share is low. (5) Administrative disputes lead to expensive adjudication. A spokesman for one operator involved in a Long Beach, California profit share contract stated that 'hassle after hassle has developed regarding charges to the net profits

*For an excellent discussion of alternative profit definitions, see McDonald, 1979, pp. 102–105.

account' (Mead, 1969). (6) Finally, because firms differ in their level of efficiency, the lessor should evaluate probable efficiencies of each competing bidder. While this is desirable, it is also expensive and may be politically unacceptable. This means that the company offering the highest profit share will be awarded the lease, even though that firm may be incapable of producing the most income for the lessor.

Under the profit share system, the incentive for an operator to manage efficiently will be severely reduced. If an 80 percent profit share bid is paired with a 46 percent corporate income tax, the reward for efficiency to the lessee firm is reduced to only 10.8 cents for each dollar saved.

Pure work commitment bidding

This system asks a bidder to specify his drilling and development program in advance. The presumed advantage of this system is that it permits the lessor to generate a higher level of exploration or development activity than would occur in response to private economic incentives. Its faults are serious. (1) It is questionable whether government officials who must approve such bids are in a position to make optimal selections. If they select any program other than the most efficient one, economic rents will be reduced. (2) It requires advance determination of an exploration or development program that often cannot be optimally determined in advance. For example, the optimal number of wells for any lease can be determined only as exploration progresses and develops information on the structure and its reserves, if any. (3) There is no objective test of the high bid. This fact may lead to corruption of government officials. (4) As with profit share leasing, the work program must be policed to ensure performance. This leads to unnecessary costs and reduction of economic rent.

CONCLUSIONS

The system used for leasing oil and gas resources on public lands should be consistent with the objective of maximizing the economic rent inherent in these resources. We have reviewed the efficiency incentives implicit in several alternative leasing methods and concluded that the pure cash bonus system is superior in terms of economic efficiency. However, it is an empirical question whether this system actually allows the government to capture the full amount of economic rent associated with oil and gas leases. We have analyzed the US experience with 1,223 Gulf of Mexico oil and gas leases issued over the years 1954-1969 by a cash bonus bidding system with a fixed ($16\frac{2}{3}$ percent) royalty rate. Our findings give strong support for the hypothesis that competition has been effective in transferring the full aggregate economic rent to the government.

Further analysis of the bidding record shows that (1) big firms have no 'unfair' advantages over small firms, (2) joint bidding is an effective means for risk spreading and information pooling and has no anti-competitive implications, (3) owners of adjacent tracts have a significant information advantage when bidding for drainage leases and this leads to superior economic returns, and (4) the larger the number of bidders, the higher the winning bid and the lower the lessee's rate of return.

We also conducted a multiple regression analysis of high bids which showed the bidding process to be economically rational. By several measures of lease quality, high bids were observed to rise as lease quality rose, while high bids fell as potential costs of development increased, *ceteris paribus*.

Based on our findings, the essential policy recommendation is that governments should choose a market oriented leasing system, the pure cash bonus bidding system, which both maximizes economic efficiency and government rent collection. Furthermore, governments should be concerned with rent capture in the aggregate and not attempt to impose rules and restrictions on particular bidders or lease categories. These can only lead to reactions from bidders and lessees which will offset the effects intended by policymakers, and may even lead to a lower level of economic efficiency and rent capture.

REFERENCES

W. J. Baumol, 'Contestable Markets: An Uprising in the Theory of Industry Structure', *American Economic Review* 72, 1-15 (March 1982).

S. L. McDonald, *The Leasing of Federal Lands for Fossil Fuels Production*, Baltimore: The Johns Hopkins University Press, 1979.

W. J. Mead, 'Federal Public Lands Leasing Policies', *Quarterly of the Colorado School of Mines* 64, 181-214 (October 1969).

S. M. Wilcox, 'Joint Venture Bidding and Entry in the Market for Offshore Petroleum Leases', Ph.D. Dissertation, UCSB, March, 1975.

SECTION 2

THE RESPONSE OF GOVERNMENT AND THE PRIVATE SECTOR

PART VII

The Role of Government

Chapter 10

United States Energy Policy

*James B. Edwards**

One of the strengths of the Reagan Administration—I don't mean this as a political endorsement—is that we have a philosophy, and we have a direction. Not everyone agrees with our reliance on the free market and our policy of reducing government involvement in the economy, but at least the Reagan Administration's views are clearly stated and our performance can be measured against our standards.

Even so, the pressure of events ordinarily leaves little time to reflect upon the longer-term trends that affect my nation and that have similar effects on other nations. I want to discuss here the lessons we have learned during the last decade and then take an admittedly speculative look into the future. I think these lessons have world-wide relevance.

We are in the midst of a long, sometimes painful, transition to new forms of energy. The transition is going to be longer than many of us supposed just a few years ago. In the past few years the United States has passed its historic peak in oil consumption; some nations have not yet reached that historic level. Hence, internationally, oil will remain the most widely traded commodity for years to come.

*United States Secretary of Energy.

There are no instant alternatives to many uses of oil. Indeed, oil has many crucial advantages; and in some applications, such as transportation, there are no readily available substitutes.

In short, there are no universal energy truths awaiting discovery. The only tools are hard work, dogged devotion, prudence, creativity, and compassion. No government can wish away a serious energy supply disruption. The American Government, like the governments of all consuming countries, must make preparations to ease the burden of an oil supply disruption. But large shortages, for prolonged periods, mean there simply will not be sufficient oil to meet all our desires. Too often, critics seem to suggest that the effects of a serious crisis can be completely ameliorated by government fiat. They can't. What we can do though, by over-reacting and over-controlling the market, is make matters worse. Twice in the last decade, the American Government has done just that.

I am an optimist. We still face major challenges in energy, but I am convinced we are learning to manage our problems. I am convinced that one day we will be able to put the energy crisis behind us. On one thing we can all agree: we are leaving that dark era where fears about energy calamities threatened our way of life, resulting in a serious loss of confidence about the future, a loss of confidence in technology, a loss of confidence in man's ability to wisely harness nature, and a loss of trust among fellow human beings. We have learned to persevere.

MACRO-ECONOMIC PRIORITIES

For a decade, many attributed most of the world's serious economic problems to the actions of the Organization of Petroleum Exporting Countries. Dramatically higher oil prices have had a serious effect on our economies. Because the increases were so precipitous, the impact was greater, and especially for some of the resource-poor developing nations it has been devastating.

It is my contention, however, that in the United States and in some other countries, the energy crisis deflected our attention away from more basic causes of our economic problems. Look at Japan. No nation is more dependent on oil imports, no industrialized nation spends so much of its foreign exchange on oil, and yet no nation has weathered the last decade better. This is not the place to analyze Japan's economy; but it is increasingly clear to us that, in the United States, higher energy prices are not the sole cause of our economic problems. Inflation is not simply the result of higher oil prices. And high interest rates are not simply the result of a large demand for energy-related capital.

Ever-increasing government spending, higher taxes, the cost of regulation, and a simple lack of entrepreneural energy are among the root causes of America's lackluster economic performance.

Energy is a commodity. It is not the only commodity to go up in price during the 1970s, although it is a more vital commodity than some. The first lesson that can be drawn from the experience of the last decade is that, while quite serious, energy problems are not the world's only economic problems.

THE ROLE OF GOVERNMENT

There are other lessons as well. In both our country and the United Kingdom, we have realized that the centralization of decision-making about energy—turning over to government the whole job of managing and regulating energy—does not work. It only makes matters worse. There is a very definite government role in energy. But we have learned that to as large an extent as possible we should allow the market to work.

During the 1970s, there was a world-wide trend toward increasing government involvement in energy. Many nations created National Oil Companies, some created Departments of Energy, and nearly all countries devoted considerable effort to legislating and mandating solutions. Not all of that effort was wrong. In the United States, it was crucial to bring our energy policy functions under one umbrella. Unfortunately, we built too big an umbrella. The US Department of Energy even today—after major budget reductions totalling over 10 billion dollars—still has 17,000 employees and about 120,000 contractors. We are continuing to slim it down in preparation for merging it with our Department of Commerce.

During the 1970s, Congress and succeeding administrations built an enormously complicated regulatory edifice to manage energy prices and to allocate supply. Faced with a crisis, there were political pressures that forced the government to act. The consequence was the creation of hundreds of thousands of pages of laws and regulations—many of them well-meaning, most of them either counter-productive or overly-burdensome.

We accept the government's obligation to regulate health and safety. In the United States we don't accept the government's right to strangle the economy with excess regulation. Consequently, we have made efforts to dismantle the regulatory apparatus that was set up during the 1970s. Frankly, I think the Reagan Administration deserves a great deal of credit for the speed at which we have accomplished this dismantlement. But there was a growing realization even during the previous American administration that energy had become over-regulated.

In 1978, Congress passed the National Gas Policy Act. The original intent of that law was to deregulate gas. Unfortunately, the Bill Congress passed has resulted in 450 pages of regulations that distinguish between more than 20 kinds of gas. The result: a serious skewing of the domestic natural gas market. Some pipelines are blessed with cheap gas, some burdened by contracts calling for them to buy very expensive gas. In fact, it is all the same gas. The price should be permitted to seek its own level.

My point is that it is all too easy to wind up with more regulations, rather than fewer, unless governments take extraordinary care to avoid legislative and rule-orientated action.

There are those in Washington who wish to impose an oil import fee as a revenue measure. I personally oppose that. Not only would more revenue encourage more government spending, it would, I am convinced, inevitably lead to a new regulatory apparatus to handle all the exemptions to the tax that farmers, fishermen, hospitals, schools, state governments, utilities, etc., would ask for. And it would lead to domestic producers suddenly raising their well-head price to match the new international price. That would force the price in thousands of contracts for oil and for natural gas—which is usually pegged to the price of oil—having to be changed.

Some industrial countries seem to feel that increased government involvement in energy matters will increase supply and assure their domestic energy security. The United States has learned its lesson.

As you know, President Reagan ordered the immediate decontrol of oil pricing on 28 January 1981. Nearly everyone agrees that the decontrol 'experiment' has been a roaring success. Next legislative season, we intend to propose the accelerated decontrol of natural gas. It is not good economics to keep the price of a premium fuel artificially low. Until the price of natural gas rises to a level akin to oil and other competitive fuels, few Americans will look hard enough at conservation. Especially in the commercial sector, there is enormous room for more conservation of natural gas.

And until the price of natural gas rises to market levels, few Americans will invest in solar hot water heaters, despite the government's 40 percent tax rebate.

SECURITY IN DIVERSITY

That leads me to a third lesson of the 1970s. We all realize that it was counter-productive to advocate reliance on a single energy source or to crusade for crash programs that ignored economic reality. Nor was it intelligent to rule out any particular energy source. We need a mix of energy sources.

In the United States, there is a bumper sticker that reads: 'Split wood, not atoms'. I like to tell audiences that anybody with that slogan on their car is unlikely to have ever split wood. I have; it's hard work. That bumper sticker also exhibits a certain cultural myopia; in some parts of the developing world, wood is the main source of fuel. There is an urgent need—perhaps one of our most crucial energy needs—to find alternatives to the rapidly depleting stands of wood that threaten the livelihood of millions of people. Fossil fuel technologies and nuclear power are essential to many of these nations.

In the United States, we went though a solar crusade. Solar power has a very 'bright' future in the US. But the goal set by the previous administration of deriving 20 percent of our energy from the sun by the turn of the century was highly unrealistic. That doesn't mean, of course, that there aren't places in the American sunbelt where that is possible. That is an economic decision that utilities, governments, and other parties in that region will have to make on the basis of careful calculation.

I continue to believe that we have to split wood and atoms and everything in between.

NUCLEAR RISK AND OPPORTUNITY

There is a fourth lesson we have learned. And it is this: while the statistical estimates of risk that occur in particular sectors of the energy industry are very meaningful, they are not always what the public perceives. Nuclear power is going through a difficult period in the United States and elsewhere. But both the statistically-derived risks and the actual experience over the last four decades since Enrico Fermi and his colleagues set off the first chain reaction in Chicago ought to convince people that nuclear power has a sound safety record that far surpasses other sectors of the energy industry. Nuclear power is safer than nearly any other large-scale form of energy generation. Compare it in particular to coal, which in the US is seen as the most viable economic alternative to baseload nuclear generation.

We are laying the groundwork for a revival of the nuclear option in the United States. At present, we have 72 licensed commercial reactors in the US, and we have about 60 more under construction. Nuclear power is far from dead. Yet before any more plants are ordered, we need to streamline our licensing process, and the financial condition of the utilities needs to improve.

We are close to securing passage of high-level nuclear waste legislation. That is apparently an issue which is not of major political proportions in England, but in the United States, lack of a final repository has become the symbol of our 'inability' to safely generate and manage nuclear power. We are going to prove that we have the legal and technical clout and political courage to move ahead.

DEMAND ELASTICITY

There is a fifth lesson. We realize now that there is a tremendous elasticity between gross national product and the consumption of energy. We should all be proud of the progress we have made in conserving energy and in improving our productivity. We believe that there is even more elasticity in energy demand than we have so far seen.

I am reluctant to project figures, we have all been sobered in recent years about the difficulty we face in accurately predicting future trends. It is widely believed in the United States, however, that for at least another five years or so, there will continue to be a large market for residential retrofits. There is an enormous market for passive designs and active conservation technologies in new homes. The commercial sector has a long way to go, as does the government. As the economy improves, we expect Americans to trade in their gas-guzzlers for more energy-efficient cars—the average age of cars being six and a half years. In short, we foresee significant advances in conservation in the years ahead.

Conservation in largely a response to price. Department of Energy studies have shown that government programs have had only a minor impact on conservation, except for energy tax credits which enable homeowners and businessmen to deduct part of the cost of their investment in energy-saving devices.

Since we are convinced that the price of energy is going to continue to increase—the era of cheap energy is definitely over—I am optimistic that we will continue to use energy more efficiently.

EXPLORATION INCENTIVES

Finally, I think there is a sixth lesson. It is that we have not yet exhausted our indigenous energy supplies. Higher oil prices encourage more exploration. Contrary to some dire predictions, we will, for a time at least, maintain world-wide reserve levels. The era of giant oil field discoveries may be just about over, but the cumulative quantities of reserves remaining to be confirmed or discovered are large. Last year in the United States, we managed for the first time in many years to increase our reserves slightly because drillers had a renewed incentive to explore and develop new fields.

In 1926, when Calvin Coolidge was President of the United States, the *New York Times* published a front page article which said 'Oil supply in sight for only six years'. We are not running out of oil yet, we are just running out of cheap oil.

Who could have foreseen only a decade ago the truly incredible North Sea Oil Field development? But oil is a finite resource. It has to be shepherded wisely. And we have to develop alternatives including coal and unconventional oil resources like shale, tar sands, and heavy oil. But this is still the liquid hydrocarbon era.

A FUTURE ENERGY MIX

This leads me to the future. I want to speculate with you about the mix of energy resources on which the world will rely in the future. These are my personal views. They are not ones to which I wish to attach any particular statistics. But they are ones that, based on present realities, are tenable possibilities for the United States. It needs to be kept in mind that not all nations will follow—or should follow—the same energy path as the United States.

First, as I have just said, we will continue to rely heavily on liquid and gaseous hydrocarbons. There will be a gradual, and hopefully relatively smooth, transition from conventional oil to exotic oils and from 'natural' to synthetic oil and gas. Coal and shale will be the main feedstock for synthetic oil and gas. We have not been moving ahead as quickly as some predicted in the commercialization of synthetic fuel technologies. However, during the last decade we have developed technologies that we know will work when they are needed.

Second, baseload generated electricity will supply a greater share of America's energy in the years ahead. Internationally, as well, electricity will probably tend to supply a greater share of primary energy. Coal will be a major source of fuel. And, in the 1990s, there will be a move toward advanced, standardized fission technologies, preparatory to a major technological advance into either fission-fusion hybrids or directly to fusion. The 1980s are the decisive decade for the breakthroughs needed in fusion. For the fast breeder, this is also the decisive period.

Third, I foresee the development of dispersed cellular technologies.

Fuel cells, powered either conventionally or by hydrogen; photo-voltaic cells and conventional 'black box' solar heaters will provide a significant share of America's energy during the next century.

Whatever the precise mix of energy sources, it is abundantly clear that during the 1980s a number of technologies will either reach the proof-of-concept stage or reach technical adolescence. I will leave it up to the professional energy economists to determine just when the marketplace is likely to judge these technologies ready for commercial application.

In short, I am convinced, that while the subject of energy will keep many professionals productively employed for generations to come, we will one day reach the point where energy is no longer seen as a problem but as a source of strength; a symbol of man's ability to safely harness the forces of nature, a source of stimulating international cooperation, and a source of hope and peace throughout the world.

Chapter 11

The United Kingdom Energy Framework

*Nigel Lawson**

Unlike economic policy, energy policy clearly means very different things to different countries, depending on their individual resources and circumstances.

It means one thing to Saudi Arabia, quite another to Japan. Even within Western Europe the differences are more marked than the similarities.

Norway, for example, with much of her electricity generated by hydropower and huge resources of oil and gas, has little concern over the cost of energy and even less over security of supply, but has to be very wary of allowing too rapid a development of oil and gas to dominate and distort the rest of the economy.

Germany, with its massive dependence on imported energy, has to balance the strategic and economic risks of different levels of dependence on the Middle East and the Soviet Union.

For the United Kingdom with its own indigenous supplies of all the fossil fuels, and a highly developed and diversified economy, the pre-eminent objective must be to ensure that the vitally important energy sector functions as efficiently and effectively as possible within the context of economic policy as a whole.

***United Kingdom Secretary of State for Energy.**

There are of course a limited number of agreed common objectives in the energy field. The Venice declaration of 1980, for example, called on all seven economic summit nations to break the link between economic growth and oil consumption through conservation and by developing alternative energy sources. Although primarily motivated by consumer concerns, this declaration was welcome to many of the oil producing countries, particularly those that take a longer view. The measures agreed at Venice have subsequently been endorsed and built up both by member states of the European Community and by the International Energy Agency. The UK has shown an impressive lead in this. Over the past two years our oil use in relation to overall economic activity has fallen by 17%.

In general, as Secretary of State for Energy in the UK, I do not see *the Government's task as being to try and plan the future shape of energy production and consumption. It is not even primarily to try and balance UK demand and supply for energy. Our task is rather to set a framework which will ensure that the market operates in the energy sector with a minimum of distortion and that energy is produced and consumed efficiently.*

ENERGY PRICING

Energy pricing is one key to this approach, in relation to both production and consumption. If energy prices are set too high producers will be encouraged to invest in new capacity for which they may not be able to find a market. If energy prices are below economic levels then energy will be used wastefully and consumers will be encouraged to invest in inefficient energy intensive processes.

But what constitutes economic pricing of energy? Where there is a genuine market—as in oil—it is the price set by the market. Where there is no genuine market—as in electricity—prices will need to reflect costs of supply. Within this general concept there is clearly some room for flexibility, for example in response to the pressure of international competition on our industries. Hence the £250 million worth of concessions to industrial energy users, especially those with high load factors, in the last two Budgets.

Realistic pricing is a stimulus to efficiency. But its impact is muted because so much of the UK energy sector is composed of state-owned monopolies. How then can we improve the efficiency of the energy industries? The key lies in increasing the responsiveness of these industries to the forces of the market place.

We have made significant progress in this area over the past three years. The changes that are in prospect will further enhance the role of private enterprise and stimulate the action of market forces, thereby increasing efficiency and helping us to ensure that the supplies of fuel we need are available at the lowest practicable cost.

I shall return later to this subject. But there is something even more fundamental. This is to recognise, as Governments have not always done in the past, that for the most part energy is a traded good.

ENERGY TRADE

Primary fuels can be imported and exported. For oil, the world market is well-established, and, although international trade in gas and coal is on a much smaller scale, it is building up. There is neither need, nor particular virtue, in having domestic production equal to consumption. The key to energy policy is flexibility. We should use our ability to import or export fuels at the margin to the best advantage in the context of an ever-changing world energy scene.

In seeking to achieve this, it does not help us very much to try to guess the unguessable—namely, what UK energy consumption will be in twenty, let alone fifty, years' time—and then aim to produce this amount judiciously divided up between the primary fuel sources. We will do far better to concentrate our efforts on improving the efficiency with which energy is supplied and used, an objective that will remain valid and important whatever the future may bring. This means, among other things, that public sector energy investment decisions should in general be based not on a simple-minded attempt to match projected UK demand and supply but rather, as in the private sector, on whether the investment is likely to offer a good return on capital. If these decisions are well based then the importing and exporting of fuels will match production to consumption on an economic basis. This does not mean that we can use imports and exports of primary fuels as a simple safety valve. We cannot turn them on or off at will. But, as international trade develops, we can expect to see our energy supply industries acquiring an additional degree of flexibility in responding to changing market conditions. And at the same time the possibility of exploiting monopoly power to raise prices will be progressively reduced.

Within this overall approach, electricity poses special problems. With the development of appropriate infrastructures coal, oil and gas can be stored, or traded, to a sufficient extent to provide market disciplines and supply flexibility. This is not true of electricity. For many of its uses there are no acceptable substitutes and, except for insignificant amounts at the margin, there is no flexibility for dealing with under- or over-supply through trade.

So the electricity supply industry, unlike the coal and oil industries, has a duty to ensure that there will be sufficient plant available to meet the top end of the range of most likely demand requirements.

But even in the case of electricity supply investment, reducing costs, to help improve the efficiency of supply and hence the efficiency of the economy, is at least as important as investment to meet projected demand. In this respect the electricity supply industry differs not at all from the industries involved in the supply of primary fuels. Diversification is also of particular importance for the electricity supply industry—a direct consequence of its limited ability to import. As many have found to their cost before, there are dangers in becoming too dependent on any one source of fuel.

In this context there is increasing public interest in the renewable sources of energy. The renewables undoubtedly offer considerable potential. They may well have a key role in the energy economy of the future, and the R D & D work which the Government is sponsoring is designed to evaluate this. But at their present stage of development it is unlikely that they will be able to make a sizeable and economic contribution to energy supply this century.

So nuclear power is critical both to diversification and to reducing costs. It is significant that the CEGB, in applying for permission to build a Pressurised Water Nuclear Reactor at Sizewell, has explicitly based its case on cost and flexibility grounds. The specific decision on Sizewell has yet to be taken. But the Government believes in general that if nuclear stations can be built to time and to cost they can play a significant role in helping to keep down the price of electricity in the UK, thereby helping our manufacturing industries in particular to increase their competitiveness.

This is bound to take time. But we can see by comparing the UK with France what the potential advantages are. In this country nuclear power accounts for some 13% of all elecricity generated; in France the figure is about 40%. This has given important parts of French industry a substantial advantage so far as its energy costs are concerned, and the disparity is unlikely to diminish for some considerable time.

In terms of meeting demand the electricity supply industry has to look many years ahead. Power stations take upwards of seven years to construct and have lifetimes of approximately thirty years. So the effects of decisions made now may still be apparent in forty years' time. However unknowable the future may be, these decisions still have to be taken—and taken on the most rational basis attainable, given the electricity supply industry's need to be able to meet economically the demand that will be made of it.

By treating energy as a traded commodity we greatly reduce the need for, and importance of, projections of UK demand and production. But for two reasons we still need to make such projections. First, for reasons I have given, the electricity industry has a special need to match capacity with likely demand and the Government which provides the finance for the industry's capital investment programmes needs to take an independent view of that likely demand. But it is only possible to form a sensible view of likely electricity demand in the context of demand for all the fuels. Second, while projections provide an unsure basis for planning they can give rise to useful questions about the coherence of policy. It is for those two reasons that my Department is now preparing a revised set of energy projections which we expect to publish in the autumn.

PRIVATISATION

The Oil and Gas (Enterprise) Act, which has now reached the Statute Book, will enable us to establish Britoil, the greater part of the British National Oil Corporation, as an independent private sector oil company in its own right and will very significantly increase the competitive pressures to which the British Gas Corporation is subject. For the first time ever there will be the prospect of competition for the custom of all gas consumers taking over 25,000 therms a year. This will provide a competitive spur not only for the Gas Corporation but also for all the potential suppliers in the private sector. We shall shortly be introducing legislation to encourage the supply of electricity by the private sector.

Where we can neither privatise nor introduce real competition we have to do our best to simulate market disciplines. The external financing limits on the nationalised industries have a crucial role to play in this respect. We are now reinforcing control by setting clearer objectives for the state-owned industries, and wherever relevant, setting performance targets. For example, British Gas has been asked to reduce its costs—other than those represented by the purchase of gas—by 5% before next April.

We are also promoting external appraisals of the nationalised industries. Management consultants are being brought in for efficiency examinations of the Atomic Energy Authority and of British Gas. The Monopolies and Mergers Commission (MMC) report on the CEGB is already leading to much-heeded changes in the way the Board evaluates investment projects. The National Coal Board and two Area

Electricity Boards are now to be given a similar examination by the MMC.

There is of course one major energy industry which is fully subject to the disciplines of the market. For North Sea oil, where an Eighth Round of licensing has recently been announced, we have a genuinely free market approach. This is most unusual among the oil producing countries of the world, among whom the UK currently ranks Fifth. But in fact it has always been the case, under successive Governments, and our removal of BNOC's special privileges has merely reinforced this important fact.

The price of North Sea oil is determined not by Government *fiat* but in response to market forces, and North Sea producers are free to produce as much oil as they wish. Even the much maligned North Sea fiscal regime is highly price-sensitive. We have made sparing use of our powers to control depletion and the rate of production, and have concentrated our regulatory intervention on minimising wasteful losses through flaring.

But we know that our supplies of oil are limited. Sooner or later we shall have to make the transition from net oil exporter to net oil importer. The obvious question is whether Government should act to defer some of the expected surplus of the next ten years or so, to help fill the gap that may start to emerge some time in the latter half of the 1990s.

At first glance the answer may seem equally obvious; of course Government should ease the way. But behind the simple facade hides a host of complexities. Prospects for UK demand and supply, and for the world price of oil over the next ten or twenty years, are highly uncertain. We also have to consider whether action now would have any real economic justification. We have to consider whether it would damage either our more immediate prospects of general economic recovery or our longer-term objective of maximising the economic exploitation of the North Sea over time.

We have to ask ourselves whether we are really so unenterprising as not to be able to put to good use the wealth which derives from oil, whenever it arises. That wealth is substantial. Last year UKCS oil and gas production amounted to some £10 bn, or 4% of GNP, without taking any account of the offshore supplies industry that rides on its back.

During the course of this short talk I have not sought to deliver a lecture on something known as energy policy. Rather I have tried to explain how I see the development of our vitally important energy resources and our energy market fitting within the wider economic objectives that the Government has set itself. This approach differs from that of previous administrations. It is one that many people with an interest in energy have perhaps been slow to understand.

Chapter 12

Achieving Structural Change: The Policies Required

*Dr Ulf Lantzke**

The present calm in the oil market has permitted a fuller debate on the proper role for the public and private sectors in energy decision-making. This is a useful—indeed necessary—debate as the energy scene changes constantly. We are now faced with examining the relevance of our energy objectives, the way we carry them out and the meaning of our achievements. Whether we are major consumers or producers of energy, and many countries are both, we cannot afford to sit back and expect events to conform to our thinking.

At its simplest, the current policy discussion boils down to whether one should let the market place determine energy policy, with governments only intervening in situations of major crisis which are an immediate threat to the national and international economy. This suggests that energy policy can be reduced to elements of basic economics with the supply and demand of energy eventually establishing a natural clearing level. This debate often is characterized by its nearly philosophical nature, whereas in practice, to my knowledge, there is no example, in industrialized countries of the theoretical alternative between total market determination of energy policy and full government control. Governments, even those with a very pronounced commitment to market economies, have always been

*The Executive Director, International Energy Agency.

aware of the importance of energy in the economy. Thus, they have attempted to influence energy policy in a number of ways. The range of state intervention is very broad in western industrialized countries, most of which are essentially market-oriented.

ACHIEVING A BALANCE

Rather than concentrating on an ideal unlikely to be achieved even were it to be desirable, we should endeavor to find the right balance between public and private responsibilities in the energy decision-making process. This, in my view, entails using the market place as the principle guiding mechanism. However, there must be sufficient public participation to keep a balance between the interests of the private investor and those of the nation as a whole.

There are several reasons for this.

First, private sector decisions tend to reflect a given company's natural interest in maximizing profits or minimizing losses over the short term. By their very nature, private firms cannot—and indeed as a matter of principle, should not—be expected to take into account the broader objectives of the economy as a whole.

Second, governments have responsibility for issues which cut across many sectors, such as defense, social welfare and general economic prosperity. They have responsibility for the smooth functioning of the economy as a whole over the long term. As such their time horizons for public action are—or at least should be—longer and the spectrum of considerations to be taken into account is much broader. Furthermore, governments do have the responsibility to resolve societal conflicts which the market has not or cannot resolve.

Third, energy policy, which affects every sector of the economy, is composed of a wide spectrum of different, and sometimes contradictory, actions. There are a variety of social, political and economic objectives which, by themselves, hinder the full realization of market forces. These constraints result from legitimate conflict in democratic societies. The market place cannot decide issues such as nuclear safety or environmental quality standards. These are areas where legitimate public institutions, including governments, alone can take final decisions. As a result, they inevitably place reins on, or to say it in a more popular way, set the framework for, the free functioning of the market.

A fourth factor which argues for some government involvement in energy is the slow reaction of energy markets to new conditions. Fundamental change taking place in energy markets is rarely visible immediately. Consequently, the response to fundamental change—

in particular the provision of new resources to re-establish market equilibrium—is slow. The response lag is probably ten years at present. In some sectors, such as electricity generation, it may take even longer. Clearly, the market will bring about the adjustments in due course. But can we, with all the political and social strains to which modern economies are subject, accept adjustment periods of ten years or longer?

A final factor I would cite is the possible inconsistency of market signals. If the signals given by the market are not clear enough, or do not remain consistent over a fairly long period, the individual consumer and investor can find himself thoroughly confused. An example, though perhaps a bit superficial, is the plight of the American automobile industry. Should it build fuel efficient cars or should it try to satisfy historical comfort standards? If the design engineers use gasoline prices as their guide, they will be changing production lines on a regular basis.

For all these reasons, I would suggest that there is a clear rationale for some government participation in energy markets. Their role is, first, to remove constraints to the full operation of the market if it is politically feasible. Second, it is also to take action of a supporting nature if the market is imperfect or even to take corrective action if objectives with higher priority are at stake.

UNCERTAINTY AND CONFUSION

If there are two words which can be used to describe developments in the energy market over the last year, they are the words 'uncertainty' and 'confusion'. I do not wish to dwell on the concept of 'uncertainty' as you are all quite familiar with what is meant here. Most of the uncertainty over the past decade has been political in origin. Few can claim to have foreseen the events in revolutionary Iran before they occurred. Fewer still foresaw the outbreak of the conflict between Iran and Iraq.

The concept of 'confusion' is an appropriate one to discuss here. We, ourselves, are to a certain degree responsible for much of the confusion that exists about developments—past, present and future—in the energy field. It is no wonder the general public is confused when it is confronted with so many conflicting opinions, all from serious, hard-working analysts. Let me give a few examples.

One group of analysts has been arguing that there is no internal cohesion within OPEC and that the resulting strains will bring about the collapse of OPEC, possibly very soon. Whether preceding this development or concomitant to it, the public is advised that the price of oil will plunge to twenty dollars a barrel or even below. One is left with the impression that the energy problem will vanish with the welcome arrival of plentiful oil at cheap prices.

A second group argues that the real solution to our energy problems lies in a massive shift from the classic energy forms to renewable forms of energy, principally solar, biomass and wind. These analysts give the impression that present standards of living can be maintained, and increased in the developing world, by reliance on these renewables, with the added benefit of little or no environmental cost.

A third group, to which the IEA belongs, suggests that world economic growth, at least in this century, will require increased supplies of coal, nuclear energy, gas and added efforts at conservation. This group argues that the link between oil use and economic growth must be broken if we are to protect our economies from energy developments such as occurred in 1973/74 and 1979/80. It argues that, if economic growth resumes at desirable levels, oil supplies will be increasingly tight unless steps are taken now to develop alternative energy sources.

What is the public to make of all these conflicting analyses? The media cannot be expected to sort out the theoretical, or even philosophical, arguments underlying these different conclusions. We must discipline ourselves to be both intellectually rigorous in our analyses and politically astute in our recommendations. Greater efforts must be made to really inform the general public.

If someone new to the energy business were to read the analyses of just the last few years, that persons could legitimately ask whether we know our field, so much have the analyses fluctuated between energy glut and scarcity. Forecasting the future by simple extrapolation from the past inevitably means that analysts are heavily influenced by current events.

All of us in the energy field have a responsibility not to confuse the situation by instant analysis. We would be wise to be skeptical of the wisdom of the day and to be loath to follow the latest fashion in the forecasting business. If I seem a bit harsh on this point, it is because I expect the analyses put out by the International Energy Agency to be judged by the same standard. We will be publishing our World Energy Outlook next October, and I would have you bear these remarks in mind when you read it. I would particularly request that judgment not be based purely on the numbers themselves, but rather on the trends that the numbers show and on the policy options suggested by these trends. For the real value of the forecasting exercise is its use as a tool to develop policy options.

POLICY OPTIONS FOR THE FUTURE

If one accepts that there is a role for government in determining energy policy and in setting the frame for investment decision by the private sector, then one must see what adjustments are necessary in the light of oil market developments over the past year.

When IEA Ministers met last month, they agreed that recent movements in world oil prices, as well as bringing welcome relief from inflationary pressures, provide an important opportunity to revitalize the world economy. However, long term energy prospects confirm the need for a policy designed to sustain and improve upon progress made in reducing the growth of total energy requirements and dependence on oil.

IEA oil requirements (including bunkers) are now lower than they were in 1973 and fell by about 4.8 mbd between 1979 and 1981 and are still falling. Total energy requirements declined in 1980 and are estimated to have fallen again in 1981, notwithstanding real economic growth (admittedly at low rates) in both of these years. From 1973–1980:

- production of energy has increased by the equivalent of 6.3 mbd of oil. Coal accounted for 37% of the total increase and nuclear power for 28%;
- the energy used to produce a unit of Gross Domestic Product in IEA economies has decreased by over 12%;
- oil use per unit of GDP has fallen by almost 19%, reflecting both greater efficiency and substitution of other fuels.

We are just beginning to make progress but the results to date indicate that our policy direction is essentially correct. Rapid increases in oil prices since 1973, the market reaction to these price increases and government policies have all contributed to reduce dependence on a single energy source, most of it imported.

However, we must be mindful of the high economic cost we are paying for sharply reduced oil consumption. The higher prices have contributed to inflation, higher interest rates, sustained low growth and massive increases in unemployment. Over the last three years, OECD unemployment has increased 11 million, reaching an estimated 30 million by the end of this year. OECD real income is estimated to be US$ 1000 billion less than if the 1979/80 oil price increase had not taken place. These economic costs indicate the social and political dimension of adjustment problems forced on industrialized countries by oil price shocks of the 1970s.

SPECIFIC ACTION

We must be ever mindful that there could be a turnaround in overall oil demand on short notice when the present stock drawdown ends or if economic activity picks up strongly. In addition, there is always the risk that political disturbance in important oil-producing regions will cause the price of oil to rise significantly. The present situation in the oil market should be viewed as a temporary one which does not necessarily indicate probable future developments. Thus, despite the progress we have made in changing the way energy is used in our economies, we must press forward with policies to sustain and improve upon our past performance.

I do not propose to go through the whole catalogue of actions which should be taken to accelerate the development of indigenous oil, coal, nuclear, gas and renewables. You are all familiar with them. Indeed, many of you have contributed to the identification of the actions needed.

I would like, however, to discuss briefly a few areas where further analysis is urgently needed.

The first area concerns the interplay between the overall economy and energy developments. Much has been written about the decline in oil consumption since 1979. However, no one—and here I include the IEA—really has a good understanding of the causes or the nature of the decline in oil use and, in particular, on the nature of the oil price increase on overall economic performance and vice versa. Analytical work is needed to assess how much of the decline has been cyclical and thus temporary and how much structural and thus likely to be long-lasting. We need a better understanding of what motivates individuals, be they large or small consumers of energy, to change their patterns of energy use. For example, what criteria do industrialists use when deciding to switch from oil to another fuel? How do we encourage them to even think about the energy component of industrial processes?

Other aspects of this issue concern the impact of technological change on energy use, inter-factor substitution, the link between growth and investment and some of the aspects of the role of market forces in affecting energy use patterns. Of particular interest, and where our ignorance is almost complete, are changing energy patterns in developing countries. To assess accurately global energy developments, we must also begin to understand what is happening in the substitution of conventional fuels for traditional fuels in developing countries. Likewise, we should have more knowledge of the impact of disturbances in energy supply on the overall economic performance which in turn has a further effect on energy demand. I believe that we don't even know yet all the parts in this inter-relationship.

A second topic which warrants serious study is the analysis and evaluation of short-term developments in the energy market, but particularly the oil market. This is clearly related to the first topic I mentioned though the focus is more narrow. The change in quarterly oil consumption is too readily attributed to the success of long-term measures. We need better analysis to determine what change is due to temporary, *ad hoc* actions, such as oil stock movements, the weather, unemployment increases, etc., and what reflects fundamental underlying trends.

A third priority area for analysis concerns electricity. Greater attention will have to be paid to the role of electricity in achieving structural change. I have a suspicion that, were one to look closely at development of technological innovation in the twentieth century, one would see a link with the spread of electrification. If this is true, we should examine how electricity penetration in industry, the household and the transport sector could be facilitated. On the supply side, an assessment of the use of competing fuels such as coal, oil, nuclear and hydro power for electricity generation should be made. We must examine the factors which may constrain fuel-switching and thus the achievement of the most cost-effective patterns of electricity generation.

Over the past decade, we have seen the energy problem strain our economies to an extent which has tested the social consensus on which Western democracies are built. We all agree that, in the future, energy should be a factor which promotes rather than constrains economic activity and human welfare. This is possible if the public and private sector pursue sound energy policies to reduce the disproportionate share of oil in our economies.

Chapter 13

The Response of the European Community

M. Carpentier*

INTRODUCTION

Given the paucity of its own resources and the pressures on world energy supplies, the European Community has a vital interest in sustained efforts to continue to increase the efficiency of its energy use and to restructure the pattern of energy demand in a manner consistent with more diversified energy supplies.

In 1973 the present ten Members of the European Community consumed and imported about 12 million barrels a day of oil. In 1981 they consumed about 9½ million barrels a day and imported (net) about 7 mbd. Over the same period gross primary energy demand fell by 3-4% while industrial output grew by some 16% in real terms.

In 1981 oil represented only a little over 50% of the Community's primary energy consumption, and imported oil some 38%. In 1981 net energy imports as a whole were down to less than 50% of our total demand—for the first time since 1965.

This is all very satisfactory—as far as it goes. But we should not be misled into inaction. There are three main reasons for caution. *Firstly, despite the fall in net oil imports the Community remains the single largest oil importer in the world (the USA imported less*

*Director General of the Commission of the European Communities.

than 6 mbd in 1981 and Japan around 5 mbd). Nearly four-fifths of our crude oil imports came from OPEC countries in 1981 and some 60% from the Near and Middle East.

The Community's net oil import bill was around $100 billion (thousand million) last year, equivalent to some 4% of our combined GNP. We remain therefore very vulnerable to the effect of oil supply disruptions and of oil price movements.

Secondly, there are practical limits to the development of the Community's indigenous supplies. Much Community coal is deep-mined and expensive to produce. The oil resources in the North Sea are large but they are not likely to provide more than the equivalent of 25% of the Community's requirements in the next few years unless there are further falls in oil demand and an acceleration of production. There certainly remains a great deal of natural gas in the Netherlands and in the British and Danish sectors of the North Sea. But many observers believe that Community production has already passed its peak. That leaves nuclear power. Its rapid development is constrained by political and social attitudes; by the long lead-times of investment projects; and by the simple fact that it can only be as electricity.

All this means that the Community is going to have to increase—and perhaps double or more than double—its imports of conventional fuels (coal and natural gas) as oil imports stabilise or decline. It means that in the transition away from oil the Community's perspective is bound to be an outward-looking one.

The *third* reason for caution is that none of us can satisfactorily explain why both energy and oil demand have fallen so significantly in the past few years, and especially since 1979.

Like it or not, governments are bound to be involved in the energy sector— whether to limit or to regulate the market power of individual utilities or companies; to tax excess profits; to protect supplies; in ensuring security and the application of safeguards in the nuclear sector; or to balance wider social interests against the interests of individual companies. And given that the effect of decisions in the energy field can take years to be felt, governments are bound to be involved in a certain amount of 'planning'—or if I could use a less emotive, though laborious term—in 'trying to make the future more manageable'.

It is quite misleading to pose the political choices in stark terms as 'intervention' versus 'the market'. But we cannot disguise the fact that there are differences of judgment between governments about where to strike the balance between intervention and the market; about the form that intervention should take; and about the inherent limitations to market forces in this rather special sector of the economy.

THE ROLE OF THE EUROPEAN COMMUNITY

The Treaty of Rome, with its particular emphasis on free trade and the progressive elimination of obstacles to competition, is concerned to a large extent with the proper functioning of the market. And in the energy sector the European Commission was insisting on market solutions, and especially on the need for realistic energy prices, long before these policies become fashionable with governments.

But the Commission is concerned not just with national markets but with the wider Community market and with the market beyond. It wants to see the implementation of a consistent approach to energy pricing throughout the Community and by the Community's main trading partners so that all of us have similar market signals and distortions to trade can be avoided.

For the same reasons the Commission would like to see common norms and standards applying to the use of energy.

But the role of the Community is rather broader than that.

In the *first* place the Community's institutions can offer a wider and more dispassionate view of developments on the energy markets than individual governments are always able to do.

I should not like to suggest that the European Commission has a better record of foretelling the future than Member States. But we may be in a better position than they are to see the likely effects of current trends and the need for individual and collective action.

Secondly, the Community institutions are well placed to see the benefits and disbenefits for the Community as a whole of interventions in the market by individual governments. We can see where they are pulling against each other rather than reinforcing each other.

Thirdly, we can hope to see where resources—human, physical and financial—are being wasted because of duplication of effort or of a failure to benefit from each other's experience. This is especially true in the fields of research and development and technological demonstration.

Fourthly, the Community as such has a vitally important role to play in external relations.

For oil, coal, natural gas and nuclear fuels, Member States need to be able to operate within a framework of relations which ensures stable, secure and economic supplies. Individually they do not have the size or the influence to secure such a framework. Collectively, they have a much better chance.

Member States have already seen the benefits of coordination within the Community. It has been demonstrated in the preparation of the Western Economic Summits; in the International Energy Agency; and, for example, in the more specialised discussions that took place in Nairobi last year at the United Nations Conference on New and Renewable Energies, and in its follow-up in Rome this year.

Fifthly, and finally, I believe that the Community as such has a particularly important role in standing up for the energy *user*. Individual governments are used to dealing with energy supply companies and all the large concerns have highly developed units for handling government relations. Of course, larger energy users also have their lobbies. But it is much more difficult for governments either to be responsive to or to influence the behaviour of the thousands, indeed millions of individual consumers and investors who have to take decisions on their energy supplies and the way they use them. For many of them the energy markets seem particularly imperfect and uncertain.

The European Commission and the European Parliament have devoted a good deal of time and effort in trying to find ways of improving the environment in which their decisions must be taken—whether by the provision of better information, by pricing policy, by improving the match between investor and potential sources of finance or by making that finance more attractive.

All governments recognise the importance of this work. And I would not claim that we have a monopoly of wisdom or that we have found magic solutions. But here again the Community dimension is important in providing a wider perspect of experience and practice.

* * *

Each of these elements in the Community role forms part of a strategy for energy in the Community developed by the European Commission and presented to, and broadly endorsed by, European Parliament and the Council of Ministers last autumn. It is a strategy that emphasises the importance of collective discipline, equality of effort and greater consistency among Member States in their policies to diversify away from imported oil and to encourage more efficient energy use. It focusses on the necessity to observe common pricing rules and principles, the stimulation of adequate levels of investment in the energy sector, a more coordinated approach to energy research, development and demonstration; measures to improve the stability of the energy (and particularly oil) market and the development of a common line in the external relations.

The three treaties—the Coal and Steel Community, EURATOM and the Treaty of Rome—provide the legislative framework for that strategy. The Community's general budget and the various financial instruments available to the Community provide the essential financial support.

The role of the Community's general budget is for the moment relatively small. But the Community's lending instruments have been playing a major role in the energy sector for a long time. In 1980 alone some £1,000 million was lent by the European Investment Bank, the instruments of the Coal and Steel Community, the new Community instrument and EURATOM together for investments in the energy sector; and energy was by far the largest single area of involvement by the Community's lending instruments, accounting for some 45% of total lending commitments.

RELATIONS WITH THE OIL PRODUCERS AND THE NON-OIL DEVELOPING WORLD

Earlier I spoke in general terms about the importance of the external aspects of Community energy policy.

Before concluding, however, I should like to say a little bit more about two specific issues which are of particular importance to the Community in its external relations.

The first is our attitude towards the oil-producing countries.

There is, I fear, a temptation on the part of some people in the industrialised world to gloat over the difficulties which the oil-producing countries have faced in the past few months and to say: 'serve you right—that's what comes of trying to buck the market'. Those who take this attitude believe that there is no longer any need to try to improve our mutual understanding.

I find this attitude myopic and misguided.

It ignores all that past history should have taught us about how quickly a buyers' market can return to a sellers' market in the oil sector. And it ignores all the uncertainties about oil supply and demand in the longer term.

Neither the oil-producers nor the oil-consuming countries have the 'upper hand' any more. This may be just the time to see what can be done to improve the meeting of minds between us. And the European Community may have a particularly useful role to play.

My final point is about the non-oil developing countries. The problems are well-known and I shall not dwell upon them. Let me simply emphasise that it is in the interest of the whole of the world to ensure that the potential energy constraint on growth in developing countries can be removed.

Developing countries are likely to be an increasingly important element in world trade in the years to come and our own economic prosperity will depend mainly on their growth and development.

How can we help?

In the first place, through our efforts to reduce our own oil consumption and to stabilise the oil markets. Developing countries will find it less easy than we do to shift to other fuels. They will be even more seriously affected by further damaging leaps in oil prices.

In the second place, we can help by increasing our efforts to stimulate domestic energy production in the developing world and to improve the efficiency with which energy is used today. This means increasing the access to new technologies and adapting them to use in different environments. It also means, quite simply to help with financing.

The Community and its Member States have a good track record in the field of energy aid. Together they are the largest source of total aid for energy investment in the developing countries, after the World Bank (committing some $1000 million in 1980 alone) and they are the single largest source of aid in the form of grants. In 1980 the European Investment Bank alone committed close to $300 million for energy investment in developing countries, helping to finance purchases of equipment, construction of energy supply facilities and so on. The European Commission itself has been heavily involved in critical areas of technical aid, especially in the field of energy programming (helping to develop supply and demand balances and thereby to identify rational policy options).

We have already done a good deal. But we can do a great deal more. And we can do it better as a Community by coordinating more closely amongst ourselves the help that we give; by ensuring closer cooperation with the financial institutions of the oil-producing states; by liaising closely with the multilateral institutions—notably the World Bank and the UNDP (United Nations Development Programme); and by greater bilateral and multilateral cooperation with our industrial partners.

Public sector intervention seems to me to be unavoidable to help mobilise the necessary investment finance, especially for heavy capital investment projects and for projects with long pay-back periods or involving new technologies. But there must also be an increased private sector effort. The developing countries have their own responsibilities in this respect in ensuring that the conditions are right for the encouragement of both public and private investment from outside. And at the end of the day it is they who must control their own destinies through sensible energy planning and management.

PART VIII

The Response of the Business Sector

Abstracts (see Section 5)

Chapter 14

Changes in Trade and Investment

*John Mitchell**

STRUCTURE OF TRADE

The structure of trade has changed and is still changing. In all the major oil exporting countries investment, production and pricing policies—and practically all the profits—are now clearly in the public sector. Partly as a result, and partly as a consequence of increasing public intervention in some importing countries, there has also been an increase in the so-called government-to-government deals: more precisely there has been a significant proportion of oil trade covered by contracts between exporting and importing state corporations or their agents.

This may or may not be important. What is undoubtedly important is that there has been a dramatic shortening of the life expectancy of crude export contracts. Over the past ten years the framework has changed from the norm concessional structure of twenty or fifty year equity production and export rights to one in which long term means anything over a year; 'long term' contracts typically have three-month break clauses, and there have been times when a lot of trade has been done on an even shorter term basis. Another

*Head of Policy Review Unit, British Petroleum.

characteristic of the new system is the absence of a norm: there is a very wide diversity of prices and terms.

These changes in trading structure have, in my opinion, had three main consequences:

1. *The number of companies engaged in crude oil trading has increased.* In the turbulent markets of 1979-80 it was possible for individuals and small firms who took risks in acquiring small quantities of oil to make very large profits. They often used them to expand their oil trading activities. Recently, they have contracted out of the market with amazing speed. I think small independent crude oil traders now have permanent access to the crude oil trading schene, just as they have long been a part of the bulk product marketing scheme in Europe and offshore the US. It is characteristic of such traders that they come and go with the opportunities. The rest of us are still there and my point is simply that the composition of the private sector has become more diverse. There is room for argument about whether the fluctuating participation of such traders increases the efficiency of markets and whether it is stabilised or destabilised as a result. I doubt whether there is an answer from first principles. I think it might be useful to look at the exercise of other commodity markets and its economic theory, to try to identify the *conditions* under which the activities of speculative traders promote or damage efficiency, and stabilise or destabilise short term markets.
2. *Vertical integration is no longer the dominant form in the international oil industry.* There are of course large private sector companies which have a spread of purchasing, trading, refining and marketing activities and therefore have some risk-spreading financial benefits of integration even although every transaction by every part of the company is at arm's-length open-market prices. When profitability shifts between different segments, as it is doing now, average profitability is less disturbed. The major state oil producing corporations do not have that advantage though they are increasingly moving into bulk product trading either through expert refiners or processing abroad. One of the questions for the future is how far the state corporations will go downstream and go abroad to do so.
3. *Some oil exporting countries are also actual or potential gas exporting countries. Where both oil and gas are in the public sector there is a possibility of their governments linking development and trade policies in the two commodities.* With some uncertainty about the share of the energy market between oil

and gas in importing countries, such exporters have some diversification of risk and possibly some strategic advantages which are not open to the private sector or the consumer side: most gas importation is in the hands of utilities who are not also oil importers (though there are important exceptions, such as the Japanese electric power companies).

I have referred to the state oil corporations and I think it is right to mention them in the same breath as the private sector because they are also part of the business sector. I do not believe that the oil market is about to be carved up into a honeycomb of bilateral deals. One reason why this is likely to be difficult is that importing government agencies cannot guarantee consumption except of relatively small fractions of the trade. That difficulty is likely to be permanent. Another difficulty is that, for the time being at least, the memory of 1979–80 undermines confidence in some exporting state corporations' ability to write credible long term price and availability contracts which will not be upset by their own governments. This may change as memories fade and more governments stand back from day-to-day market management of their national oil corporation.

In the meantime, state corporations have to operate in the market with the private sector. Recent events show that in the international market, state ownership does not necessarily guarantee commercial success in the international market. The National Oil Corporations have to trade rather like the private sector corporations if they are to trade successfully.

As with any other commodity in international trade, the market might be made more efficient if it were easier to predict intervention by governments either directly or in the commercial activities of private and public sector corporations. Inter-governmental agreements could theoretically offer increasing predictability. There are of course certain sovereign rights which governments will never agree to restrict. But as an optimist I wonder whether there are not some aspects of the oil market which might some time be candidates for multi-lateral governmental agreement. There are after all other international commodity trades in which state corporations and private corporations are major participants and in which inter-governmental agreements on the framework for trade have been reached from time to time. GATT is one example, commodity agreements are another. Neither model is obviously applicable to the oil market and I would certainly be sceptical about any ambitious attempts to determine by governmental agreement the course of future oil prices.

In short, as far as responses to the changing structure of the oil market are concerned, I think that the private sector participation has become more diverse, and is adapting to a new structure in which public sector corporations have an important part. I believe the public sector companies in the international market will increasingly accommodate to the disciplines of the market place, and that it is worth thinking about whether their task, as well as that of the private sector, could be more efficiently performed if any part of the pattern of direct government interventions in the international trade could be stabilised.

OUTLOOK FOR INVESTMENT

Conventional wisdom admits the possibility of disruption and of serious economic damage as a result of disruption to the oil market. I guess conventional wisdom does not exclude the effect of another such disruption on the longer term balance of demand and supply and therefore price.

My version of one of today's new conventional wisdoms about the energy future is therefore that it is possible that for the next seven to ten years there may be no strong trends in the prices of internationally traded oil, gas and coal, though there may be wide short term fluctuations. This view can be applied both to absolute levels of the price for each fuel and to the relationship between prices of different fuels.

The scenario of energy prices which fluctuate randomly but show no convincing trends is not the only one a prudent planner should carry in his wallet, but it is the only one I will discuss today. It is having some influence on the expectations of the private sector.

An essential part of the story is that, though the paths for total energy development may be relatively smooth, the outlook for any individual fuel will be very uncertain because the rate of substitution for oil, and the timing of projects for substitute supplies, are very unpredictable. It is this uncertainty about timing which makes long term plans for any particular fuel so difficult to formulate either in the private sector or in government.

If a sufficiently large number of investments for new energy supply for the 1990s really depend on expectations of profits from significantly higher prices, they may be delayed until the delay itself generates the expectation of higher prices on which they depend. The energy industry may face a pattern of capacity development, familiar in the minerals industries, which cycles between boom and bust. The uncertainty created by the capacity cycle will increase the risk premium into the returns required by investors in the industry.

Many of the investors are private sector investors, either directly in energy companies or indirectly in the financial markets in which most state corporations have ultimately to finance (however indirectly) their capital projects. Wise men, and perhaps an equal number of economists, may ask whether the animal spirits of the market are the best available regulator of capacity development. Anyone could also ask whether all of the present regimes of public intervention in energy investment are necessarily economically efficient or politically stable. I am sure there are no general answers to either question.

Let me highlight a few particular problems in the following paragraphs.

One of the few certain things about the future is the inevitability of the decline in production from almost all the thirty large oil fields which were responsible for 40% of the increase in oil supplies between 1950 and 1973. In very round numbers, it is almost inevitable that the fields now producing 22 mbd outside OPEC will be producing not more than half that amount or less in the year 2000, and that within OPEC the present production capability of between 30 mbd and 35 mbd will decline by the year 2000 to 20-25 mbd if there is no further development and exploration. Just to maintain an oil production capability, outside the Communist areas, of around 50-55 mbd therefore requires new or additional capacity over the next twenty years of 10-12 mbd outside OPEC, and about the same amount within OPEC.

Though such development requires new discoveries, especially outside OPEC, I believe that such discoveries are possible. They will not be made if there is no expectation of development, and development depends on economics and in turn on the animal expectation of businessmen. I remind you that by necessity there are business instincts in state corporations as well as in the private sector.

As you know, the geographic distribution of significant oil reserves is highly concentrated. Outside OPEC, probably 80-90% of the potential for new or replacement production capacity lies in five countries. One of these is Mexico. The others are the US, Canada, Norway and Britain. I think it is probably fair to say that about half of the remaining opportunities world-wide for old-fashioned private sector investment in oil production are concentrated on these four countries. In two of them, Norway and Canada, a large role has been designated for the government corporations. I will not discuss the past work or future role of state corporations in the development of capacity in the UK.

The obvious point is that the future of the classic private sector production activity is limited and is highly dependent on the specifics of leasing, depletion, tax and pricing policies in a very small number of countries. In some of these, policies towards the private sector seem at times to be based on the absolute confidence that ever-rising oil prices will cover the industry's ever-rising costs so generously as to leave the government with ever-rising oil tax revenues.

I hope and expect that the star actors of the private sector exploration and production industry will always have a few stages on which to play their classic roles. It is obvious, however, that there are other shows in town. In most of them a wide variety of private sector companies display technology, organisational strength and the willingness to risk front-end money. They will continue to do this in return for the prospects of rewards which are correlated as strongly with skill and luck as with the mere ability to spend money. On this basis private sector companies of all sizes, and with a variety of national origins, have struck mutually acceptable bargains with governments, and with state corporations, in many countries both within OPEC and outside it. In fact, there has been an increase in the number of countries in which private sector oil companies are engaged.

Brazil, India and China are important examples of countries where private sector companies are doing or are about to do work in a framework where the incentive is preserved for them while the major share of the profit, and ownership and policy for exploitation of the resource, remains consistent with the national political structure. That is all good news about flexibility and adaptation and I expect that to continue. The bad news has to do with nearer home.

Outside OPEC, the new capacity development I have described may be mainly offshore, under conditions of increasing cost. With an outlook in which there is no clear trend in oil prices, taxation policies become critical. The relative attractiveness of the private sector role in these countries can deteriorate faster than legislators can learn about the oil business. Capacity development outside OPEC may be slower as a result.

The oil supply curve in the 1990s will therefore be the product of two market factors. Only one of these is power of the governments of the dominant exporters. The other is policies of the governments of the four leading OECD producers towards the oil companies who will be responsible for much of OECD oil capacity development. In the non-Communist world at present, private sector oil companies are probably responsible for about two-thirds of total investment in oil exploration and production, and they are trying to finance it essentially on the cash flow from under a third of total production.

The oil companies therefore represent a wider business sector. The 'untrended outlook' means that oil development projects are looked at within the alternative opportunities in the world economy. Even major oil companies which do not diversify very far from oil ultimately resort to the financial markets of major OECD countries. Public sector companies are in the same position as long as their governments face fiscal deficits at the margin. If policies for economic recovery in general lead to improved corporate profitability in general, and a revival of investment in the economy at large, then the 'untrended scenario' implies that the attractiveness of oil investment compared to other investments will weaken. I fear that after tax profitability of oil investment over the next ten years is more likely to be squeezed than enhanced, as costs rise faster than prices while the tax systems lag behind developments in the real world.

What about the business role in the development of other energy supplies for international trade? In the case of coal, the major increases in supply are expected from private sector companies. LNG schemes typically involve many partners in one or more segments of the operation. Nuclear development is mainly in the hands of public sector corporations or regulated utilities. I repeat one general point about the effect of the 'untrended scenario'. Investments for energy supply are investments in the production and delivery of a particular fuel, and the rate of growth in demand for any particular fuel is more uncertain than the rate of growth in the demand for energy in general. Private sector companies, unlike public sector corporations and utilities, can diversify the timing risk by diversifying across oil, coal, gas and perhaps uranium. For the diversification to be effective, however, the possibility of periods of above 'normal' profits in one fuel must be permitted to offset the possibility of simultaneously below 'normal' profits in another. Tax and policy approaches based on narrow concepts of 'rent capture' frustrate this diversificiation. The result is higher risk premiums and consequently higher costs for consumers in general and slower growth of the competitive fuels share of the energy markets.

In short, the private sector has an important part in the development of new oil and other energy supplies for the long term. The rate of development will be influenced by expectations of profit relative to other opportunities, ultimately in the world economy at large. The possibility of an 'untrended scenario' implies a higher risk for new energy supply projects than in the past, and will bear particularly hard on oil projects in the major OECD countries where tax regimes are based on other assumptions.

PART IX

Specific Strategies for Supply Disruption

Abstracts (see Section 5)

Chapter 15

Some Rational Strategies for Supply Disruption

*Philip K. Verleger Jr**

Oil producers, oil consumers, oil companies, and governments of consuming countries all play a part in making an interruption a catastrophe instead of a minor inconvenience. The catastrophe is, however, avoidable by prompt and effective governmental action.

Oil prices can be expected to increase by very large magnitudes during future supply interruptions despite the efforts of consuming countries to establish special programs through agencies such as the International Energy Agency. The research shows that these programs and these agencies will fail because little effort has been made to understand the nature of an interruption in the supply of oil, or the behavior of oil markets during disruptions. Specifically, a loss in supply of oil creates an initial shortage, or gap between supply and demand. This shortage quickly begins to increase as panic spreads through the market. The increase starts a process which soon develops a momentum of its own which can ultimately cause adjustments in prices far in excess of those required to rebalance supply and demand under the new, post-disruption conditions. The word which most

*Booz, Allen & Hamilton Inc.
†This research will be summarized in a forthcoming book entitled *Oil Markets in Turmoil: An Economic Analysis* (Ballinger Press, 1982).

accurately describes this process of bidding up prices is 'lethargy', not 'greed'.

Generally it is found that the market participants (consumers, companies, governments, producers) adjust slowly to changes in market conditions. This lethargic adjustment transforms minor incidents into major crises, and would transform a major crisis into a catastrophe.

While present policies are probably incapable of dealing with a disruption, rational and effective measures can be suggested once the characteristics of market adjustment are recognized. In addition, the perversity of other irrational policies such as price controls can also be assessed.

Throughout the study, the primary focus of the analysis is on the spot market for petroleum products—commonly referred to as the Rotterdam market. It is argued that this market is a barometer of conditions on world oil markets, rising during periods of shortage and falling during periods of surplus. It is shown that this market sets the world price of oil (as it should, because it is the arena in which marginal supplies are bought and sold).

The spot market also offers a means of measuring the effectiveness of various programs imposed at times of disruptions. Successful programs will stop the process of price increases. There is an exact analogy to actions taken by central bankers in defending currency. A successful defense will stop foreign exchange rates from falling, just as successful energy measures will stop prices from rising.

The movement in spot prices during disruptions is traced to the size of the shortage. The greater the shortage, the larger the price increase which accompanies it. The analysis also shows that the increase in spot prices continues as long as the shortage persists. Thus, delays in response to a shortage will drive prices higher. This characteristic has been recognized by many and has caused some to argue that controls should be placed on the spot market. Proposals of this sort are viewed here as analogous to hanging the messenger for delivering news of defeat.

My analysis of the dynamics of crises shows that the sources of problems in oil markets during disruptions are not those commonly listed by critics of the oil industry. For instance, OPEC is guilty of lethargy, not greed. OPEC uses the spot market for petroleum products to set crude prices but tends to follow that market very slowly. This slow adjustment creates an incentive for those having access to low-priced OPEC crude to profit from the temporary difference between the price set by exporting countries and spot market prices. This characteristic tends to prolong the disruption and increase the magnitude of the price increase. *Thus, one conclusion of the study is that the problems of disruptions could be dealt with more easily by raising prices quickly.* Optimally this would be accomplished by imposing a tariff, but a decidedly second-best solution would be for OPEC to raise prices quickly.

A second finding of the study is that oil companies tend to raise consumer prices too slowly during disruptions. In particular, changes in consumer prices tend to follow changes in OPEC prices. Thus, to the extent that OPEC is culpable for its lethargy, oil companies are culpable for following OPEC rather than the spot market in setting prices. Some or all of this guilt must be shared by governmental policies which have prevented oil companies from raising prices. Even where no government restrictions exist, company officials can be excused for not raising prices out of a justifiable fear of the public wrath.

A third finding is that the consumer is a contributor to the process of increasing prices. This contribution occurs through the slow consumer response to higher prices demonstrated by the difference between short- and long-run price elasticities of demand. By responding slowly to changes in prices, consumers create conditions which tend to force prices even higher. Prices are driven higher yet when price controls or other considerations delay the increase in consumer prices.

The slow adjustment in prices and the unwillingness or inability to increase consumer prices also contributes to the ultimate panic that accompanies a disruption by making it profitable to increase inventories during a disruption. It becomes possible to buy at low prices during the initial days of a disruption and then sell at much higher prices by the end. Quite naturally, this circumstance causes inventory demand to increase, further worsening the crisis.

RECOMMENDATION ONE: RAISE PRICES QUICKLY

Interruptions in the supply of oil can be dealt with most easily by aggressively raising prices, because higher prices start the conservation process which is required to return supply and demand to equilibrium. Further, a quick increase in prices will prevent oil prices from 'ratcheting' to even higher levels by the end of the crisis. Since OPEC countries have been reluctant to raise prices quickly in the past, the governments and oil companies in consuming countries should be willing to initiate the price increase.

RECOMMENDATION TWO: RAISE PRICES BY IMPOSING A LARGE TARIFF ON THE IMPORTATION OF OIL

A tariff on imported oil will force up consumer prices and induce quick conservation. It will also reward those who have built up speculative inventories of crude oil and product by allowing the holders of these stocks to sell them for the higher prices (inclusive of tariff) without sharing the windfall with the government. This feature will tend to dampen the increase in spot prices and thus benefit consumers through lower long-run oil prices.

A tariff will also reward those who develop capacity to produce incremental supplies of oil or other fuels to the extent that these gains are not captured by taxes.

RECOMMENDATION THREE: PROGRAMS SHOULD BE ADOPTED TO ENCOURAGE THE DEVELOPMENT OF GREATER PRIVATE STOCKPILES

The impact of a disruption on prices depends on the level of private stockpiles. The greater the stockpiles, the smaller the impact on prices. Thus the adoption of measures which either reduce the cost of holding stocks (such as tax subsidies) or increase the expected profits from holding stocks in the event of a disruption should be considered.

RECOMMENDATION FOUR: CONSUMING COUNTRIES CAN CONTROL THE RATE OF INVENTORY DRAWDOWN AND DEFEND ANY OIL PRICE THEY SELECT THROUGH THE USE OF A GRADUATED TARIFF ON OIL IMPORTS

In the long run, consumers must cut consumption to match any reduction in supply. The way in which the cut is managed will determine the rate at which oil prices rise. I would advocate the use of a graduated tariff which starts at a high value and then is quickly cut back over several months. A graduated tariff would:

- Cause a maximum reduction in consumption by overcoming the inertia which characterizes consumer demand

- Cause a maximum rate of stock sales in the early months of a disruption by offering the maximum profit to speculators who sell early
- Minimize the draw on strategic stocks, since the cut in consumption and sales from private stocks will partially relieve the shortage
- Encourage consumers to postpone consumption by holding out the promise of declining rather than rising prices.

A graduated disruption tariff can be used to defend any oil price in the same sense that a central bank defends the value of a currency. My analyses show that the volume of stocks used during the early months of the crisis will depend upon the size of the tariff and the price that is defended.

RECOMMENDATION FIVE: MECHANISMS FOR RECYCLING THE RECEIPTS OF THE TARIFF ARE ESSENTIAL

Many studies have shown that an interruption in oil supplies would have serious consequences for the US economy due to the increase in the price of oil and the transfer of wealth from the US to producing countries. The imposition of a disruption tariff would block this transfer of wealth, but the macroeconomic impact of a disruption can be minimized only if mechanisms which allow the instantaneous rebate of tariff proceeds are put in place before the disruption. Incremental proceeds from the Windfall Profit Tax resulting from the increase in prices must also be rebated if fiscal drag is to be avoided.

RECOMMENDATION SIX: REFINER AND CONSUMER TAXES ARE NOT SUBSTITUTES FOR A TARIFF

Some have suggested that taxes on gasoline or refiner output be imposed in lieu of a tariff. These proposals negate the potential benefit of speculation. Specifically, the imposition of a tax on a product denies the speculator the windfall from the liquidation of stocks. Thus, if standby taxes are enacted rather than a tariff, speculators will hold fewer stocks and tend to husband them longer during a disruption. Both actions will magnify the price increase which accompanies a disruption.

RECOMMENDATION SEVEN: ACCESS TO PUBLICLY OWNED STOCKPILES SHOULD BE EASY AND QUICK

One of the most significant responses to the Arab oil embargo was the decision to develop strategic oil stockpiles in many consuming countries. Given my findings about the effect of stockpiles in meeting disruption, this decision was clearly correct. However, these stockpiles are of no use unless the oil is accessible quickly at the start of a disruption. Thus, some mechanism must be established to enable consumers (not governments) to control the timing and the rate of withdrawal. This is particularly important since publicly owned stockpiles tend to reduce the size of privately owned stocks.

RECOMMENDATION EIGHT: EMERGENCY CONSERVATION MEASURES WOULD BE HELPFUL IF IMPOSED QUICKLY, BUT WITH FEW EXCEPTIONS, CANNOT BE PREPLANNED

Emergency conservation measures represent a very useful element in any program to meet a disruption. However, to be effective, the programs must be imposed quickly and must make real reductions in consumption. Unfortunately, one finding of this study is that the conservation measures which appeared promising in 1973 or 1979 either did not work or worked so well that they have now been fully incorporated into daily life and thus offer little further potential for conservation. In a future disruption, the only measures which are guaranteed to work are the politically unpopular ones such as banning weekend driving.

RECOMMENDATION NINE: PRICE CONTROLS AND ALLOCATIONS ARE A TERRIBLE MISTAKE

The standard response to a disruption has been to impose price and allocation controls. No action could be more detrimental to the long-run interest of consumers and consuming countries, because controls delay adjustment and drive prices up even higher. Further, the provision of allocations or determination that some companies or some consumers require special treatment removes any incentive for these companies and consumers to build precautionary stocks and thus reduces the world's stockholdings, increasing the probable price effect of any disruption.

RECOMMENDATION TEN: INTERNATIONAL COOPERATION IS A GOOD IDEA BUT NOT ESSENTIAL TO MEETING A DISRUPTION; FURTHER, THE PRESENT STRUCTURE OF INTERNATIONAL COOPERATION AS SET UP IN THE INTERNATIONAL ENERGY AGENCY IS COUNTERPRODUCTIVE

The analysis developed in this study assumes that measures to meet a disruption are imposed uniformly by all developed countries. However, it is not necessary that all countries respond in the same fashion, but only that the response is quick, calms the spot market, and gets consumption down to the new level of supply before stocks are exhausted.

While international cooperation is helpful, and perhaps even essential in meeting a disruption, coordination through the International Energy Agency as it is presently constituted will probably prove to be a disastrous mistake for two reasons. First, the IEA as an international consultative body cannot move quickly. Diplomats and bureaucrats from the member states will need to meet to discuss an ongoing disruption. While those discussions take place, consuming countries will be reluctant to take independent actions for fear of giving up something at the bargaining table (as happened in 1979). In the meantime, prices will spiral ever upward and panic will develop. This could, of course, be overcome by giving the IEA the authority to unilaterally take action at the start of a disruption. However, it is unlikely that the member states would assign it such authority.

The second problem with the present IEA arrangement is its misplaced focus on quantity. All IEA emergency plans address shortages of varying magnitudes. Yet interruptions are not primarily a problem of quantity but of price. It is the increase in price and the transfer of wealth that create the great economic dislocations, not the shortages. Rational energy policy can and should be designed to achieve the sudden reductions in consumption necessary to meet a loss in supply without allowing prices to rise indefinitely.

BIBLIOGRAPHY

Energy Action, 'Statement of Edwin Rothschild, Director, Energy Action

Educational Foundation, before the Subcommittee on Fossil and Synthetic Fuels Committee', September 9, 1981 memo.

National Petroleum Council, *Emergency Preparedness of Interruption of Petroleum Imports into the United States*, The NPC, Washington, DC, 1981.

Plummer, J. 'Methods for Measuring the Oil Import Reduction Premium and the Oil Stockpile Premium', *The Energy Journal* 2 (No. 1), (January 1981).

US Department of Energy, *Reducing U.S. Oil Vulnerability, Energy Policy for the 1980's*, US GPO, Washington, DC, 1980.

SECTION 3

ENERGY IN WORLD POWER POLITICS

PART X

The Growing Impact of Soviet Energy

Chapter 16

Communist Bloc Energy Supply and Demand†

*A. F. G. Scanlan**

Let me begin by saying that whatever the definition of Communist Bloc may be in other contexts, the essential 'core' of countries in an energy context is the USSR and the six Eastern European full members of CMEA: Bulgaria, Czechoslovakia, East Germany, Hungary, Poland and Romania. Mainland China is a separate story altogether and has no energy trade with the Bloc since Romanian oil ceased to arrive at Shanghai as oil self-sufficiency came to China and left Romania. Some of the other Communist countries, especially in Asia, are more significant—up to half a million barrels a day has to be allowed for them out of the Soviet oil balance—but that is as far as it is necessary to note them until one of them discovers some oil.

Siberia dominates the CMEA energy picture. All future increases in CMEA oil and gas production potential will have to come from Asiatic USSR. Although Eastern Europe retains significant coal potential, much of it is low grade; inside USSR the potential for expanding coal output also lies in Asia. The size of the deposits—Tyumen, oil, Ob Delta gas, Kuznetsk and Kansk-Achinsk coal—gives significant economies of scale in the actual unit cost of production

*Public Affairs and Information Department, British Petroleum Company.

†See also Chapter 19: Why Europe Needs Siberian Gas.

but the total cost to the economy of infrastructure, transportation and importing European manpower into almost totally uninhabited regions makes unit costs a residual rather than a determinant consideration.

USSR

Oil

Production doubled from 6 mbd to 12 mbd in the past decade solely due to a successful development of West Siberian oil, which accounted for all the increase. Some measure of the dominance of Tyumen oil is that five out of every six barrels produced in the USSR come from either West Siberia or Volga-Urals but production in the latter area has declined steadily since 1975. As a result of West Siberian production—7 mbd out of the All-Union 12 mbd—the USSR exports 3 mbd which means it is exceeded only by Saudi Arabia in the league of world oil exporting countries. And of course the USSR leads the world oil production table. However, this means that nearly 5 billion barrels is depleted every year.

Internal CMEA demand absorbs 11 out of every 12 barrels produced in USSR, 9 in USSR, 1.5 in Eastern Europe and 0.5 in other Communist countries as already noted. Eastern Europe is the region that probably has the largest oil deficit in the world—more of that later—but even if one makes the most generous assumptions about internal USSR demand, it is hard to see other than a steady increase in the transport and agricultural sectors which account for about 3 mbd at present and for which no effective substitutes exist.

The reality is that a considerable momentum in the economy is needed to *generate* the Asiatic energy supply—perhaps as much as 2% of 'simulated GNP'—in order to keep the five year plan moving. The other problem with the notion that economic slow-down will alleviate the oil or energy balance is that there is little evidence to support such a relationship in the Soviet economy. Production targets are met, absorbing energy in the process, while the total economy responds more slowly than industrial production itself. Consumers in western economies lose purchasing power and buy less energy in a period of slow-down, but in CMEA unemployment is disallowed and wages are not the sole arbiter of consumption since disposable income may actually be in excess of the accessible supply of goods and services. Allocation, rather than price or cash flows, determines economic activity and this is true in fuel choices too: the customer neither pays an economic price for using energy nor chooses which primary source heats his apartment. That is done centrally in both senses.

Natural Gas

The outstanding success in discovering and developing a major internal CMEA natural gas network has been going on now for over 20 years, during which production has increased ten-fold. The gas reserves in the USSR are about 40% of the entire world reserves, mainly in the Ob Delta, Orenburg Oblast and Central Asia. All these areas are linked to the main consuming centres. I recall comparing the network achieved in the 1971-75 Plan for all large diameter pipe as equivalent to a Trans-Alaska Pipeline every eight weeks throughout the Plan. The point to note today is that that comment was made nine years ago since when other major additions—SOYUZ to complement BRATSVO, for example, into Eastern Europe—have been concluded. It is not a development at the grassroots stage!

USSR exports of gas to the rest of CMEA have exceeded oil exports in incremental volume since 1975. In other words, gas has become the major export growth fuel into Eastern Europe as well. But it is the internal Soviet scene that is being overlooked in the debate that is currently surrounding this issue. My NATO paper last year drew upon Soviet sources to indicate how gas is replacing oil in all the leading energy intensive industrial processes except oil refining. And in the domestic sector, gas connections have been effected at twice the rate of new apartment buildings, i.e. massive retrofitting has been in progress to older buildings for a decade—so that just before he died, the Gas Minister, Orudzhev, claimed that 200 million people are now connected to the gas grid. That broadly equates with the entire urban population of the USSR.

It is clear that between 1975 and 1985 the major elements in a structural shift from oil to gas will have been made in both industry and domestic sectors in the USSR. After 1985 diminishing returns and spatial logistic problems outside the cities will set in with a vengeance. Oil is under pressure especially in Eastern Europe—so the obvious solution with surplus gas reserves, a saturated internal gas market in key sectors and a need for hard currency—is to export the gas. On a smaller scale this can be seen in Far East developments with Japan—but essentially it means extending the pipeline network from Eastern to Western Europe.

All this is well-known, but presented in this perspective it suggests that, short of burning more gas in power generation, the export outlet is the *only major one* for gas as the decade progresses. Exports to Eastern Europe already indicate gas growing faster than oil, the opposite to what is required there. Gas production is scheduled to increase by over 40% in the current five year plan and by 1985 will equal oil in the energy balance. The major export increment is likely to emerge after that date, later than envisaged officially, but dovetailing closely in timing with recent hints in Pravda or delays in field drilling and gathering systems and with declining growth in demand in Western Europe so that the new Urengoi/Yamal export total is now set at 35 billion cubic metres rather than 40 bcm and may end up no more than 30 bcm in the event. After 1985, Soviet forecasts indicate that the next increase in gas supply will be less than is planned for coal, although the gas reserve base could undoubtedly provide more. This is probably because Soviet policy is *not* to burn gas (except locally) in more and more power stations, a point often made by leading figures such as Minister of Power, P. Neporozhniy. The last oil-fired power station has been built in USSR and the future lies with coal, hydro and nuclear.

Coal and Electricity

Three-quarters of the reserves are in Asia where the more accessible western parts around Kuznetsk and Karaganda (northern Kazakhstan) were developed for deep mining to augment the Ukraine and Moscow coalfields. The Donetz is now producing 10% less than its productive peak of about 225 mta a few years ago and now lower grade opencast coals east of Kuznetsk at Kansk-Achinsk and at Ekibastuz near Karaganda are being developed for use in local power stations. Asiatic coal has taken over the major part of coal production, but growth has been insufficient to do more than offset the decline in European USSR so that all-Union production has remained static for five years.

The lack of investment priority in deep mines cannot be rectified within the decade so hopes for the planned increase from 700 million to 775 million tons by 1985 rest upon opencast 'soft' coals in Asiatic USSR and their use locally or transmission by 'wire'. The decision to built power stations at the coalface has now committed the USSR Power and Electricity Ministry to this option rather than expanded railroads or slurry pipelines, assuming the latter were feasible. The rail option, with nearly half of all rail freight traffic already represented by coal and its requirements, would severely tax a system still giving priority to the second line to the Pacific, the BAM. Thus even if the coal targets are met everything depends upon long distance transmission for which new high voltage direct current (DC) lines are being developed. These lose less power through resistance over long distances although they have to be converted into alternating current (AC) and the use of converters on such a large scale is a technical frontier at which equipment delays or failures could be crucial. It is noticeable that dates for commissioning these lines have receded twice during the current five year plan. Neither the coal production targets for 1985 nor the All-Union Grid link-up is likely to be effective until the next Plan period when coal expansion expects to benefit from an investment priority it has not had for 25 years. But the increase expected in the period 1986–90 is about three times the increase forecast for 1985 which you may recall is the 1980 figure slipped five years. A similar fate awaits most of the post-1985 planned increase in terms of timing.

Because of the difficulties of 'transmitting' Asiatic coal, European parts of Comecon including the USSR are scheduled for a most ambitious nuclear power station programme. Output is planned to double within six or seven years. While this is well within the technical capacity of CMEA, we know now that Atommash will only add eight or nine new reactors in the Plan period, so that the timescale is clearly at risk of major slippage. If this occurs, then technical progress in 'coal-by-wire' will not be a surplus for export but a vital stopgap internally.

In general terms, it may be a reasonable middle view of Soviet coal prospects that they will increase very little, perhaps 10%, in effective contribution to the end of the decade. If this sounds pessimistic, one has to allow somewhere for the Soviet practice of quoting mined output in gross tonnage, in many cases coal containing 30–40% impurities, and for the deteriorating quality of the average gross ton as production moves away from hard coals to lignite. Over the past 20 years, the average calorific value of a ton of mined coal has declined by no less than a quarter. The incremental effect of dependence on Asiatic opencast lignite is likely to accelerate this trend. Power transmission losses and rail wagon losses over greater distances will add to the increasing gap between mined raw coal and its effective contribution to usable energy. Of course, any relief to the oil position that coal can achieve is highly valued, but the question may arise as to whether coal is to relieve the burden on oil or the burden on nuclear power.

SOVIET EXPORTS

Eastwards

The strength of the USSR energy balance is unquestionable, even if their level of oil production is forcing massive substitution. It is only when the six countries of Eastern Europe (the 'Six') are added to the balance that real problems emerge. Before looking at westward prospects in general the eastward pattern—essentially trade with Japan—

is worth attention because it brings into focus another aspect of Soviet energy, namely, the separate character of the Soviet Far East. One has to travel 4,000 miles east of Moscow to reach the Pacific, and two-thirds of the journey is eastward of the major Siberian and Central Asian energy deposits. Two thousand miles east of Tyumen in the Yakutsk ASSR gas and coal discoveries have been made too far east of any population centres to allow any sensible outlet except further eastward to Vladivostok and Japan. Offshore the oil and gas of Sakhalin offer better sited but more limited prospects for these two markets.

Japan imports about 8,000 barrels per day of Soviet oil and 1 million tons of coking coal per annum. The cities in the Pacific region such as Vladivostok are the only significant population centres in the 3 million square miles of the USSR east of Lake Baykal and are insufficient on their own to justify developments 1,500 miles northwest in Yakutia. The Japanese are therefore cast in the role of developer. So far, they have accepted commitments to develop coal at Neryungi in Yakutia which is scheduled to provide about 5 million tons of coking coal annually by rail to Japan via Nakhodka from 1985 to the end of the century: in the meantime, the coal is delivered from Kuznetsk. The Japanese have not accepted a similar commitment for Yakutsk gas, preferring a trilateral deal involving the USA as equal importer. If the climate of acceptability to all partners were to re-emerge, the project would take 10 to 15 years to develop, each importer receiving 7-8 bcm from Olga. However, the project for Sakhalin LNG, from the mainland port of De Kastri, is due to commence by 1985 at 5 bcm annually.

Important as the infrastructure development is strategically for the USSR, the quantities are tiny—about 1% in the case of Japanese oil and coal and 9% of gas needs by 1985. The vast bulk of Siberian resources are westward oriented; lying at the eastern edge of the populated area but still in the western half of the USSR.

Westwards

Virtually all Soviet exports of energy are involved with either Eastern or Western Europe. This was not always the case; in the 1960s a vigorous oil export programme included several Latin American, African and Asian countries: but Eastern Europe has tripled its imports of energy—oil and gas—from the USSR since 1968 and by the early 1970s we were speculating on how the choice between developing world, hard currency earning (Western Europe and Japan) and CMEA requirements (Eastern Europe) would be made. The developing world was first to go, although other forms of trade deal remain and several supplies of oil—not Soviet oil—emanate from these so that, for example, Iraqi and Libyan oil meet Soviet third party commitments elsewhere as rouble exports. Now the question is being asked how the final choice will be made between the two parts of Europe as gas supplies increase and oil exports face up to rising USSR demand outstripping production increases.

More recently, speculation increased on the oil problem in Eastern Europe. The USSR and Western Europe both have a strong interest in natural gas, the USSR to maintain the momentum of developing its new major fuel—it will surpass oil production in the Soviet Union within three years and make the USSR the leading world producer of gas—and of course to ease the hard currency earning burden placed on oil. Western Europe has learnt, like Japan, to diversify its considerable import needs away from oil and to diversify its sources of gas supply as much as possible.

EASTERN EUROPE

There is not space in one paper to review all the aspects of Eastern European energy. A detailed account 'East European Energy and East-West Trade in Energy' by Jonathan Stern has aptly just been published as the first of a new series of joint energy policy studies by the Royal Institute of International Affairs and the Policy Studies Institute in London in association with the BIEE. The point I wish to emphasise here is that the impact of what must ensue from the energy crisis in Eastern Europe is quite different from the impact of the long-term energy scene in the USSR.

The key features in Eastern Europe are unique for a grouping of developed nations: an almost total dependence on imports for oil, an abnormally low percentage of oil in the energy balance and therefore a low or zero ability to substitute other fuels for oil for certain key economic purposes, and membership of an economic union, CMEA, whose most substantial export capability is, paradoxically, energy. Add to that that their use of energy per capita is one of the highest in the world, especially taking into account relative economic attainments and their peculiar sensitivity to imported energy prices becomes clear.

The three northern countries, Poland, Czechoslovakia and the DDR average a mere 20% of oil in their primary energy mix and to find a comparison with the 15% of oil in the Polish energy balance one would have to go to China. The southern countries average about one-third of oil in the energy balance, similar to the USSR, but with

Romanian production now in decline prospects are worsening. The great mainstay is solid fuels, mainly low grade outside of Poland and therefore only of use internally. The combined production of solid fuels in raw tons in a normal year is, at 700 million tons, equal to USSR coal output, but the USSR, with two and a half times the population produces nine times as much gas and sixty times the combined oil output of Eastern Europe.

From 1975, since when incremental Soviet gas supplies have exceeded oil deliveries, the terms of oil imports have gone from bad to worse. The rouble price of Soviet oil was linked to a five-year moving average of world prices and quantities above plan targets were set at full world price. Before the world oil crisis in 1973 the CMEA importers had paid the Soviet Union a full arm's length price for their oil and the effect of the 1975 revisions was to double the 1973 price—less than the full effect, certainly, but bad news for CMEA importers nevertheless. During the period between the two oil crises, Eastern European imports of non-CMEA oil quadrupled from 5 mta to over 20 mta. The second oil price explosion ended that at a stroke.

The Soviet Union has reaffirmed that quantities above plan levels will be at full world parity and are imposing a collective ceiling at about 80 mta (400 million tons for the Plan period). The severe check to economic growth and the rescheduling of international loans has ensued, but the real answer to the higher real cost of imports must lie not in contraction but an expansion of export trade, either with the Soviet Union or elsewhere. By raising the price to world parity—both planned and unplanned quantities of oil from the Soviet Union will probably be awash at this level by 1095—the door is potentially being opened by the USSR for Eastern Europe to pursue this course. It is significant that world price parity is already being suggested for quantities above plan volumes in 1982. What market economist could possibly object? Many countries in the developing world, equally squeezed between high oil prices and tight money may find new evolving relationships around the not inconsiderable technology and needs of Eastern Europe. This in turn could link trilaterally to oil producers due to their own depressed circumstances, or to sheer disaffection with the existing system of trade in the current recession. Was it nature or economics that abhors a vacuum?

The Soviet attitude is not one of disengaging as has sometimes been ascribed to this line of analysis. How could it be when three-quarters of the oil and all the gas imports in Eastern Europe are dependent upon the USSR? But the increment is another matter.

WESTERN EUROPE

The proposed new gas trunkline deal would approximately double the volume of gas. Statistically this might represent 3% of Western European energy requirements that comes from CMEA—although by 1987 total consumption will have risen and Soviet oil exports may have declined somewhat.

One element in the CMEA pattern of oil exports that has already ceased to make sense is the re-export of products from crude oil purchased in hard currency. CMEA oil product imports into Western Europe are the largest single source of finished products but unless in future such trade is based on ruble crude or barter crude the Western European refining surplus is unlikely to give the trader refiner much incentive. By the time that refinery surplus is worked out ruble crude may not have attraction for Eastern European refiners if it is available only at full world price parity.

This is not the place to go over the pros and cons of West German exposure to Soviet supplies. The diffusion of the total quantity of Soviet export between several nations all interlinked in the western gas grid and with the flexibility of surge capacity, storage, interruptible contracts, interfuel substitution and international co-operation at a far higher level of readiness than in 1973 are recorded by Stern and others elsewhere. What seems to me more important than denying what, at the very least, has been described by Paul Frankel as greater 'diversification of insecurity' is a willingness by the western consumer nations to maintain their co-operation generally in this regard. The beneficial effect of any additional source of energy under normal conditions must be to increase flexibility generally and reduce the chance of pressure on prices.

The significant point that is emerging is that the main impact of changes in the Soviet supply pattern affect Eastern, not Western Europe. If by the time the Soviets plan to reach full export capacity of the Urengoi-Yamal gasfields they also plan to sell oil to *all* parties at world prices, with gas somewhat lower, they effectively pass the whole problem to their customers in Eastern Europe. They put their customers in Eastern Europe under enormous pressure. It is the *rate* of change in world oil prices that is most damaging to importing economies who tend to either over-react towards stringency or to rush for induced expansion, neither route having the structural time to work through without either inflation or stagflation resulting. If in the longer term, the squeeze on oil prices induces real economies and real productivity compensation through expansion in trade, equilibrium is re-established. Can Eastern Europe manage with a lower percentage of oil in its energy mix? I believe the answer is still no. The potential to economise on energy is large, but not to substitute oil by other fuels. For example, except in Romania, the extraordinarily high percentage of solid fuels in electricity generation—75-90% in the three northern countries—takes out at root the option the OECD countries—and the USSR—possess to back oil out of the electricity sector. Even if GNP growth is limited to 2% annually, Eastern Europe has a need to increase oil supplies at at least the same rate—another 10 to 15 million tons annually by 1986. If by the next plan period, 1986-90, the USSR is receiving a full world oil price from both Eastern and Western Europe the breakdown between the two is of secondary importance and, equally so, is the amount of gas sold in each part of Europe providing that the target quantities for both exported fuels are reached. In this regard the gas exports are a vital component, but a significant increase is feasible through the existing pipeline system.

It is unlikely, therefore, that a significant impact is due to occur either in Western Europe or the USSR as a result of changes in the composition of CMEA energy exports. The impact in Eastern Europe is much more severe, and with the decline in Romanian oil potential and the difficult road back for Polish hard coal during the current world recession, the danger of further implosion—or as in the case of Czechoslovakia stagnation—is unpleasantly real.

Further expansion of the lignite base in all the economies except Poland and the over-optimistic timing of the nuclear power programme may mean further slow-down or greater dependence on the USSR and, if possible, the OECD, to allow time for the turn-round. Transfer of certain energy-intensive activities to the USSR and extension of USSR electricity exports may help to improve the national energy coefficients in Eastern Europe.

For the USSR itself, the 1986-90 plans detailed at the ECE and in Ministerial statements, are a portent to a far greater impact at the end of the decade. If these plans are achieved the USSR would not only have become the leading oil producer and gas producer but also the premier coal producer in the world.

Greater emphasis is placed on coal expansion in the second half of this decade than on oil and gas combined. Oil is under pressure—it may have peaked during this, the previous plan, and problems with further gas penetration may be a principal factor in the relatively slow increase in gas compared with the current period. The nuclear power programme, highly concentrated in European CMEA, Russia included, will be up against the time factor, and at the base of all this development will be Asiatic coal, not European, and the crucial question of the rate of progress of 'coal by wire' in a nation spanning eleven time zones. Will the planning emphasis on coal now be matched then by increases in the offer of electricity exports to OECD Europe?

The answer to this question depends almost wholly on the answer to another—will the CMEA countries still require roughly twice the energy input of Western European countries for the same economic output? In March this year the BBC reported a Moscow radio broadcast as stating that the potential for fuel saving in CMEA was 900 million tons of standard fuel. That is the same problem, quantified. There can be little doubt of the sincerity of the Soviet policy to remain self-sufficient in each form of energy, but it must now be obvious that measures to improve the economy on the demand side of the equation have now got to augment further production increases. The position is more urgent in Eastern Europe where self-sufficiency is out of reach and economic viability itself may be the central issue.

Meanwhile the trade balance should not present any undue problems, unlike those in Eastern Europe. The Soviet pattern is regularly one in which 10% of their oil and 5% of their gas pay for imports of food, machinery and raw materials in about equal measure—essentially to assist Soviet agriculture, heavy industry and energy development. There is a major prize to be gained in improving the agriculture deficit, either in crop productivity or in improving the delivered percentage of the crop, although it is perhaps not sufficiently appreciated by the world at large that grain output has risen 50% since the end of the Kruschev period. This is outside the scope of this paper except to note that a 10–15% improvement in the delivered agricultural produce would be of as much advantage to the balance of payments as to the total current volume of gas exports. Such feedback from international trade may provide the system with the opportunity to evaluate demand management. *The main impact is not whether the Soviets achieve all their energy production targets by 1990, but rather where do they go from there unless efficiency in demand takes a decisive step forward.*

PART XI

Towards Self-sufficiency in North America

Abstracts (see Section 5)

Chapter 17

Price Reactive versus Price Active Energy Policy

*John F. O'Leary**

The focus of US energy policy and its principal strategic ingredients have been essentially unchanged since the time of the first reaction to the oil embargo of 1973. Energy policy throughout the 1970s and early 1980s has been designed with the overwhelming objective of reducing US oil imports regardless of their source. We have adopted four major strategies to achieve this paramount objective. Each of these strategies has been utilized to some degree by each of the four successive US Presidents that have been in office since the time of the embargo. Although the emphasis—the policy mix—has changed, the policy ingredients have remained the same.

The four strategies include: measures aimed at achieving reductions in use of energy; measures designed to achieve fuel switching, particularly from oil and gas to coal; measures adopted to encourage domestic production of conventional fuels; and of course, actions aimed at the creation of new energy sources primarily through research and development.

The objective of oil import reduction has been achieved to a surprising extent. Today's import level is barely half of that experienced in 1977. The import reduction, however, has occurred in ways and with associated cost not foreseen by US policymakers.

John F. O'Leary Associates, Washington, DC.

The primary contributor to import reductions has been the slowdown in economic activity we have experienced since the most recent siege of oil price increases in 1979. These reductions in economic activity have primarily affected energy intensive industries. Additional reductions have been a direct consequence of the permanent shift in economic activity in the US to less energy intensive industries. Most of the remaining reductions have resulted from price induced conservation that has forced the substitution of capital and, alarmingly, of labor for energy with a consequent reduction in the capability of the US economy to satisfy final demand.

EFFECTS OF OIL IMPORT REDUCTION POLICIES

A review of the four policy areas indicates the following results. With regard to fuel switching, despite the enactment of two major pieces of legislation and an enormous outpouring of rhetoric the efforts aimed at forcing large industrial and utility users to coal has had essentially no influence on demand for fuel in the United States. The early efforts of the Federal Energy Administration and more recently those of the Department of Energy to implement congressional policy mandating the conversion of 'coal capable' facilities from oil or gas back to coal have uniformly failed. This has occurred, first because of inadequacies in the law itself, second, because of the determined opposition of the utility industry, and finally because of the influence on policymakers of the perceived surplus of oil and gas of the last year or so.

Similarly, efforts aimed at reducing oil imports by development and promotion of new technologies have been a complete failure. Despite the fact that we have spent very large sums of money over the past thirty-five years in the pursuit of alternatives to the conventional fuels, we are not now receiving any benefit from these expenditures and it is altogether unlikely that we will be the recipient of such benefits during the remainder of the century.

There has been modest progress in the deployment of some solar based technologies particularly flat based collectors and passive heating for residential construction. These contributions are tactical, however. Moreover, they appear to be driven by psychological as much as by price considerations and it is difficult to connect their growing acceptance with any of the technology programs that have been supported by the United States Government. Similarly, in the area of conversion of coal to liquid and gaseous fuels, the little progress that is being made is proceeding on the basis of technologies other than those developed by the Government as a result of its very substantial efforts in this arena. In summary, we can conclude that in the non-nuclear arena Government R&D activities over the past thirty-five years make virtually no difference in the current energy equation in the United States.

There has been a similar lack of success in non-price related attempts to stimulate conventional production of oil and gas. Production of both fuels is above levels that would have occurred had the repressive pricing policies of the early 1970s continued intact. These output levels, however, are measurably lower than would have been the case had the Government accepted a hands-off policy respecting energy pricing. Again, the non-price policies in this area have failed to contribute to the overall goal of lessening import dependence.

It is important to point out that much of the increase in current production of both oil and gas is being secured at the expense of future production. In oil particularly the current levels of output are not supportable on the basis of current and prospective findings. Only the more optimistic analysts believe that the current levels of output of natural gas can be sustained at current and reasonably predictable finding levels.

There have been, in the United States, a number of different approaches to the problem of obtaining improved conservation of fuels. These have included direct regulatory initiatives with a punitive aspect of violators (the adoption of the universal 55 mph maximum speed limit in the United States); direct subsidies to improve the behavior of energy consumers (the weatherization program that was adopted in the mid-1970s); and, of course, there have been the tax incentive programs where good behavior in conservation terms is rewarded by tax credits or other direct or indirect subsidies. In addition we have attempted to develop standards, both mandatory and voluntary, that would guide manufacturers in the development of energy consuming appliances and would guide consumers in purchases of these articles. In retrospect, despite the enormous amount of intellectual energy expended in the United States to find alternatives to price driven conservation, these programs have simply not succeeded. As we take a look at the very real change in the relationship of energy to constant dollar GNP it appears that virtually all of it has been derived from conservation induced by price increases and from a basic change in the industrial mix of the US economy favoring less energy intensive activities. The combination of regulatory, rhetorical, and subsidy devices that have come into law in the United States over the past decade have had little effect on consumption.

HIGHER ELASTICITY OF OIL DEMAND

A review of recently available energy consumption data indicates that oil has become much more prone to sharp swings than is the overall use of energy. In 1981, for example, the world-wide recession resulted in a decline in total energy consumption of 0.6 percent from 1980. Petroleum consumption, however, declined by almost 3.5 percent on a global basis. In the US total energy consumption in 1981 was 3 percent below the previous year while petroleum consumption fell 6.4 percent, representing virtually all of the reduction in energy use. This phenomenon has been observed in most industrial countries.

It is probable that the same selectivity that forces a disproportionate drop in oil consumption during downturns in economic activity will be manifested by a comparable degree of selectivity on the upturn. In the US, for example, it is likely that return to 4 percent annual GNP growth in 1983 or 1984, should it occur, would be accompanied by a disproportionate increase in oil use—and therefore of oil imports. This is so because a return to 4 percent growth in GNP would necessarily be accompanied by a resumed high level of activity on the part of those energy intensive and particularly oil intensive industries that are now so seriously and selectively depressed. This would almost certainly be accompanied by a return to higher levels of petroleum product utilization on the part of those individuals and organizations that have retrenched in response to real or threatened economic distress.

In addition, we may find that reductions in oil consumption in the industrial countries are more apparent than real. It is altogether likely that the data available, which are apparent consumption data, fail to reflect highly significant movements in stocks held by downstream factors in the industry. There is substantial evidence that these stocks, which are largely unreported, rose sharply during the last half of 1979 and through the summer of 1980. The severe economic downturn experienced over the past year, the extraordinary high rates of interest and the disappearance of the widely held perception that the real price of oil would perpetually advance, have combined to force the elimination of a substantial portion of these stocks. There is at least anecdotal evidence that currently in the United States the unmeasured secondary and tertiary stocks are at critically low levels mirroring the position of reported product stocks.

Taken as a whole we might find that a restoration of reasonable rates of economic activity in the US could increase imports by 2 mbd to as much as 3 mbd from the present distressed level,returning imports to the levels of the mid-1970s. It may well prove to be premature for the US to claim victory in the battle to contain oil imports.

As we review the energy history of the past decade it is quite clear that our policies that have been designed primarily to confine the impacts of price increases have failed. It is equally apparent that unless there is a significant shift in our policy objectives with regard to our energy economy, we will continue to be exposed to the same sorts of price shocks during the 1980s and beyond, that we experienced with such devastating impact in the 1970s. Despite nearly a decade of applying the four strategies that I have outlined earlier, we find ourselves today no less, and in dollar terms possibly even more, exposed to the devastating effects of interruptions in oil supply than we were at the time of the embargo. Indeed, an examination of the experience of the 1970s leads to the conclusion that the overwhelming policy objective in energy of the past decade, that is to secure a reduction in oil imports, has not worked, and furthermore is not relevant for dealing with the problem with which the United States is confronted. It is becoming increasingly recognized that the crucial energy problems are those associated with long term influence of grossly inflated oil prices rather than with short term energy availability.

INCOME EFFECTS

It is, of course, impossible to quantify precisely the costs to society of the energy price escalations of the past decade. We are able, however, to obtain some understanding of these impacts as a result of analyses that have been done by a number of observers. Probably the most significant costs associated with the energy price shocks have been in their long term influence on gross national product. In the United States the current level of economic activity is below that which would have ever reasonably been expected in the absence of the events of 1973 and 1978 by $200 to $300 billion annually. This conclusion is derived from work that was published by Otto Eckstein in 1981.

The point is made all the more strongly in a 1981 report of Resources for the Future relating energy costs to family income. In that report we are reminded that in the United States family income after a long period of steady growth has not increased in real terms since the time of the embargo and has been maintained during the past decade, largely because of the movement of women into the work force. Further the Resources for the Future data tell us that direct and indirect energy expenditures which may have been in the area of 5 percent of disposable family income in the early 1970s, are now in the neighborhood of 15 percent of family income. One can conclude from these estimates that since the time of the embargo real family income net of energy outlays has actually fallen by something in the range of 10 percent. This in turn has been a major contributor to the reduction in savings that have characterized the period and the concomitant falling off of investment. These basic changes in the wellbeing of our people, of course, are having major political consequences in terms of attitudes toward Government expenditures. Indeed, it is probable that the current political upheaval in the United States is directly traceable to this phenomenon of significantly reduced real disposable income and even more important, the significiantly lowered expectations that have followed from the 1973-74 and 1978-79 energy experience.

Clearly, the enormous gains in real productivity in the industrialized world during the post-war era were a direct result of the substitution of cheap energy for expensive human labor and for capital. There is strong statistical evidence to support the thesis that much of the decline in the growth in productivity that has been observed in the United States and in other industrialized nations since 1974 has been a consequence of reverse substitution. That is, we have during the past decade made a significant substitution of labor for energy as energy prices became an increasingly significant portion of industrial and commercial budgets.

Among the most difficult analytical areas is the influence of oil price increases on inflation. While there is no particular consensus with regard to the proportion of today's inflation directly attributable to energy price shocks, there is a generally held view that the inflationary impact in the United States and indeed globally of energy price increases has been substantial.

We now find ourselves in the second major recession since the embargo. The first post-embargo recession, that of 1974-75, would according to many analysts, have been no more than a recurrence of the relatively moderate downturns that we have experienced repeatedly since the end of World War II had it not been for the oil price increases associated with the embargo. In fact that recession was the most serious in the United States since the depression of the 1930s. Similarly, the current position of the United States and of most of the industrialized world of severe recession and high unemployment is more than coincidentally related to the energy price increases that were associated with the Iranian Revolution and its oil price sequel in 1979. Again we are experiencing the most severe of the post-depression downturns. Clearly, the oil price shocks of the 1970s are by no means the sole cause of the ills of the US economy. It is probable, however, that the dominant set of incidents in the 1970s that have converted ordinary cyclical phenomena to ones of extraordinary severity have been the changes in energy prices set in motion by the two oil supply emergencies of the decade.

It is clear, in retrospect, that the economic damage that the United States sustained during the periods of oil shortage of 1973-74 and 1978-79 was relatively trivial. We found, for example, in reviewing the 1973-74 episode that the decrement in GNP during the shortage itself from what would be regarded as 'normal' economic activity was not more than $10 billion at annual rates. A number of analysts have projected the damage in post interruption period arising from the changes in price generated by the shortage as at least an order of magnitude higher than the costs to our economy during the emergency, *per se*. Similarly, an assessment of the 1978-79 oil shortage period would show that the damage in terms of reduced production of goods and services in the United States during the crisis period was small and manageable in comparison to the extremely high cost that has been paid by our economy (and those of other industrialized nations as well) in the post-emergency period. Again the ratio is at least one to ten and perhaps a good bit higher. It is probable that the cost to date of the two energy emergencies in terms of lost GNP is well over a trillion dollars for the United States economy and of course the US has borne only a portion of the world's total.

THE LEAST COST PRINCIPLE

In adopting our policies with regard to the commodities that are used in commerce, with a few minor exceptions, our attitude has been to apply the least cost principle. That has been a hallmark of our attitude with regard to energy in the United States. Going back to the earliest interventions of Government in energy markets with only a few exceptions, we have attempted to minimize energy prices. Indeed much of the social history of the United States of the 1930s and even earlier was intermixed with the creation and development of energy initiatives that were aimed at keeping availabilities high and prices low. I cite in this context the early emergence of utility regulations, the adoption of energy depletion allowance for oil and gas, leasing policy on the public domain, the creation of the Tennessee Valley Authority and the Bonneville Power Administration. Much more recently, the willingness of our policy system to continue price controls on both oil and gas for long periods of time was a further manifestation of this attitude.

The least cost principle as applied in the United States has had two distinct elements. First attempts were made to limit costs of production of fuels and of other raw materials. Oil and gas costs have been lowered as a result of the Government's active encouragement through subsidies for research and development. Public land policy until the late 1960s was designed to provide maximum stimulus to the extractive industries and further designed to minimize their costs. The US Government has for most of this century maintained comprehensive information programs aimed at cost minimization for the oil and gas and coal industries.

At the same time significant efforts have been aimed at limiting rates of return in the energy producing industries to 'reasonable' levels. The purest manifestation of this has been in electric and gas utility regulation and in natural gas field pricing, but it has applied in one form or another to all of the fuels. The oil industry, for example, has been for most of this century a notable target of antitrust efforts aimed at extracting real or imagined economic rents from the crude and product price structure. In more recent years, comparable efforts have been directed at coal, particularly at the state level.

Over the years prior to the embargo the least cost principle had become distilled into a set of policies designed to assure that our energy raw materials came to markets under terms that maximize public benefits.

In the energy debate that has raged in the United States since the embargo, we have lost sight of the least cost principle as the focus of energy policy. We have sought, awkwardly, to contain the effects of the oil price increases that have resulted from the two oil emergencies of the 1970s, but we have not taken concerted action to influence oil prices, *per se.*

Our energy policies, in particular our oil policies, should return to the least cost principle that served us so well during the earlier years of this century, and rather than simply attempting to constrain the negative effects that are associated with ungoverned price increases for oil they should be aimed at influencing oil prices over the long term to return to something in the realm of cost of production plus a reasonable rate of return—the criteria that governs the price of virtually every other commodity that is significant in international trade. The failure of the United States and of the other industrialized nations to pursue least cost as the guiding policy criteria in its dealing with the energy equation of the past decade has forced an incalculable cost on the whole of modern day society.

The enormous progress that mankind has made in material terms during this century has come about largely because of the unimpeded flow of large volumes of low cost energy. The dampening of that progress that we have noted in the United States and indeed throughout the world in the past decade has been a direct consequence of impeding that flow, not because of any basic shortage, but rather as a result of price. In assessing the effects of the increases in productivity that have not occurred since the time of the embargo, the reduction in the capacity of the US and other western economies since the embargo to produce goods and services for their own economies and for the remainder of the world at ever declining prices, the stimulus that oil price increases to levels several times above costs have provided to the inflation that we now experience on a global basis with a severity unparallelled in modern history and the continuing threat to stability in the Third World as a result of the fiscal strains imposed by oil price increases, it is clear that there is no single task that could have wider and more beneficial consequences to the global economy than to return real oil prices to levels commensurate with their cost.

PRICE POLICIES

The tools to accomplish this task are available, and have been well documented. They consist of actions under three general headings: Inventory Management; Allocation and Price Control; and Diversification Through International Cooperation.

The embargo and the Iranian Revolution both occurred at the end of long periods of liquidation of petroleum inventories. In 1973 and again in 1978, product stocks were at critically low levels, as they are today. In contrast, oil stocks in the fall of 1980, at the outbreak of the Iraqi–Iranian War, were abnormally high. The first two events led to frantic bidding to meet current demand and to replenish inventories. (Stocks at the end of both supply emergencies were higher than at the beginning.) This panic buying set off the price increases which in turn caused the long-term damage that we have experienced. High stocks prevented panic in 1980 and an event potentially as disruptive as the Iranian Revolution passed without oil price effects.

Should a comparable disruptive episode occur today, panic bidding would again almost certainly rule the market and the disastrous events initiated by the two rounds of price increases of the 1970s would be repeated.

The cure for this is relatively straightforward. Governments should be much more attentive to inventory management practices of industry and should have and employ the authority to require industry to maintain stocks at levels that would discourage panic bidding at the outset of a supply crisis.

The strategic stockpile unless it is enormous (and therefore too expensive) is not going to play much of a role at the beginning of any future supply emergency. The problem with utilizing it is that there is an overwhelming tendency to hold it until an emergency clearly exists. In 1979 we rapidly learned the magnitude of the oil lost from Iran but we could not know the duration of the loss. We had 100 million barrels in the strategic reserve but we kept it for a 'true' emergency. In the meantime the price went up and up.

The strategic stockpile should be utilized during the course of an emergency at the moment that it is clear that prices will be forced up in the absence of such use. It must be, above all, constructed and managed as a price influencing device.

Price controls and allocation go hand in hand in managing a supply crisis. Fundamentally, there are two ways to allocate goods and services. The first, and almost always preferable technique, is to permit supply and demand to be equated by price.

The reasons why this is not a good idea for oil during supply emergencies are quite straightforward. First, oil demand is extremely inelastic in the short run. Over one or two years, a price increase for oil can make a real difference in its use pattern. For a shorter period, say six months, our experience has been that we can have a major change in price with only the most minor changes in consumption. If price is relied upon to allocate a shortage during an emergency, in order to reduce demand, say 10 percent, it may be necessary for prices to increase 100 percent. For almost any other commodity that is acceptable because the price will return to a level representing basic resource cost plus a reasonable rate of return once the emergency is over. But for oil we have learned that, whereas OPEC has not demonstrated an ability to raise real prices, it has mastered the technique of following panic bidding-driven increases in the spot crude market with step increases in Official Selling Prices (OSP), and in holding OSP in periods of slack demand following the emergency. (The past six months have provided an object lesson in this holding capability.)

The political dynamics of oil pricing dictate that once the industrialized nations have shown a willingness to pay high prices in the spot market, the Governments of the producing states, to satisfy their electorates, must capture the new high price in the OSP. Once captured, it is of course tantamount to national disgrace to lower the official price. Thus, unlike other commodities, once oil prices rise in today's environment, only the attrition of inflation can cause them to fall.

If, in consideration of the above, we decide to assert price controls, then it is necessary that formal or informal allocation come into the equation to accomplish the market balancing function ordinarily performed by price.

All of this is familiar ground. We have two bitter experiences with allocation and price controls in the United States. Bitter because we had all the pain of allocation—lines, profiteering, uncertainty—and prices at the end of both experiences were enormously higher than at the beginning.

If our Government is going to use price controls and allocation again it must find means of assuring that its citizens get something for their injury. That something is an assurance that prices will not advance.

THE ROLE OF THE INTERNATIONAL AGENCIES

The prevention of price increases during supply emergencies necessarily requires relatively broad international support of the sort that the International Energy Agency was devised to provide. The Agency, however, has never had a sufficiently strong charter to permit it to do what must be done to prevent further rounds of oil price increases. What must happen is that the IEA should become the harmonizing agency for development in the principal industrial countries (and among them critically Japan, Germany and the US) of a set of inventory, allocation and, above all, pricing policies that will prevent further rounds of panic bidding. The US, Japan and Germany have dominated the spot market in the two past emergencies, and have led the increases in that market that in turn have been followed with increases in OSP. There are two points to note here. The first is that the vigorous bidding in the spot market did not increase the net amount of oil available to the bidders, rather it influenced each bidder's share of a rigidly fixed total. Thus painful price controls and allocation internally were frustrated and made largely pointless by unconstrained bidding for product and crude on the world market. The second point is that the supplying nations did not anticipate or expect the price increases experienced in the supply crises of the 1970s. There was at the time of the embargo a price goal for OPEC of roughly $5.60 per barrel, just half of the price that finally resulted from the bidding-up of the spot market that occurred in the first half of 1974. The formal OPEC price goal at the time of the Iranian Revolution was simply to retain the embargo-secured gains in terms of real purchasing power, which would translate to roughly $20.00 per barrel today.

During the 1979 emergency we utilized a regulatory tool that could be of major assistance in maintaining price stability. Many companies in the first half of 1979 were bringing into the US oil that had been acquired at high prices in the spot market as well as oil in which they had an equitable interest and which was bought at lower Official Selling Prices. The Department of Energy handled this problem for price control purposes through the so-called 'Deemed Equity Rule'. Under this regulation, if Company A had crude available from Country Y on an equity basis at an OSP of say $20.00 per barrel, and purchased spot crude originating in the same country at $30.00 per barrel, the spot crude was 'deemed' to be equity crude and credited at $20.00 per barrel into that company's crude price mix which in turn governed the price it was permitted to charge for gasoline and other products.

The philosophy behind this rule, which was fully sanctioned by law, was that a company was free to pay whatever it chose for oil on the world market, and of course, supplying countries were free to charge whatever they wished. But for price control purposes, the US was equally free to recognize only those costs that it chose to recognize in administering its domestic price control program.

I have no doubt that, had we had the foresight to extend the concept of the Deemed Equity Rule to all of the crude entering the US, with maximum prices reflecting OSPs at the beginning of the crisis, and had we been joined by Japan and Germany in a comparable effort, all or virtually all of the increases in the spot market that occurred in 1979 could have been avoided, and the consequent ruinous increases in OSP avoided as well.

Finally, the much more difficult set of problems associated with diversifying energy supplies to de-emphasize the Persian Gulf must be addressed.

Implementation of the World Bank's initiative, developed under Robert McNamara, recommending establishment of a special program aimed at aiding Third World countries to meet a substantial portion of their energy requirements would be the first step in this diversification program. Beyond that two other broad initiatives would be required. The McNamara idea should be expanded to include other potential energy projects that would increase the amount of energy, particularly oil, in the global trade pool, e.g. Venezuelan heavy oil. And the US synthetic fuels program should be restored to its 1922 objective of 2 million barrels daily of new capacity.

The problems presented by the accident of nature that placed a totally disproportionate share of the world's cheapest source of energy in the Persian Gulf region are manageable. We have now spent a decade in efforts to come to grips with the problem, and sadly we are no more, and in sheer dollar terms perhaps less, able to deal with the situation than we were in 1973. In the meantime, First World economies flounder, Third World hopes are dimmed, and the entire world continues for the indefinite future to live with an ever-present threat to peace.

At this point the world clearly has two courses open to it. We can continue to subject ourselves to the terrible dynamics of oil pricing and look toward further dissolution of the international and national fabric that developed to the benefit of all of us during the 1950s and 1960s or we can adopt the alternative course of bringing the world into a renewed era of stability and economic growth insofar as energy policies can contribute to that objective, by adopting a set of policies designed to restore stability to energy prices.

PART XII

Pricing Problems in the Pacific Basin

Abstracts (see Section 5)

Chapter 18

Market Price and Resource Cost in Australia

*G. D. McColl and D. R. Gallagher**

INTRODUCTION

Australia's position in the world economy appears to have been strengthened as a result of the international energy situation which has arisen in the last decade, mainly because of its relatively large endowment of coal, natural gas, uranium and bauxite. The main purpose of this paper is to review the current energy situation in Australia and to consider the projections which have been made for the current decade and beyond.

There are clear indications that the changes in energy consumption which have been occurring in other OECD countries since the second round of oil price increases in 1979 are also in evidence in Australia. Between 1977-78 and 1980-81, sales of petroleum products declined by nearly 5%, consumption of black coal rose by more than 16% and natural gas by about 50%. As a consequence, the share of petroleum products in total primary energy consumption fell to less than 40% in 1980-81, compared with 48% in 1970-71. The contribution of

*Centre for Applied Economic Research, University of New South Wales. The authors are indebted to Thomas Mozina for research assistance in preparing this paper. The full paper is in the BIEE Archive.

black coal increased to nearly 32%, brown coal continued to contribute about 10% and the share of natural gas rose to 12.5%, compared with only 3.4% in 1970–71 (Joint Coal Board, 1982, Table 98).

The structure of energy demand and supply in Australia illustrates a number of well-known features of Australia's energy situation:

(a) Reliance on petroleum products for about 55% of total final demand for energy, approximately two-thirds of these products being consumed in transport uses and nearly one-half in road transport;
(b) Net imports of petroleum, principally crude oil, are equal to about one-third of domestic consumption of liquid fuel;
(c) The heavy dependence of electricity generation on inputs of black coal, particularly in New South Wales, Queensland, South Australia and Western Australia, and on brown coal in Victoria;
(d) Significant use of black coal in manufacturing activities, particularly in the iron and steel industry;
(e) Substantial exports of black coal from Queensland and New South Wales, equal to about two-thirds of total coal output;
(f) The use of natural gas in manufacturing, and, to a lesser extent, in commercial and domestic uses, now amounting to about 12% of final primary energy demand;
(g) The export of uranium from recent mining developments, principally in the Northern Territory.

PROJECTIONS OF ENERGY USE

Projections made by the Department of National Development and Energy imply the following changes during this decade:

(a) An increase of approximately 120% in total primary energy output, of which only about one-fifth would be required to meet increased domestic demand, the remainder being accounted for by a trebling of coal exports, an even faster increase in uranium exports and sales of LNG abroad equivalent to 7.9 mtoe;
(b) A very small increase in the quantity of petroleum products consumed, with further reductions in their consumption in manufacturing, and as a consequence, a fall to less than 32% in their overall share of energy consumption;
(c) An increase of about 80% in electricity production, accompanied by similar rates of increase in the inputs of black and

brown coal, together with further strong increases in the use of black coal in manufacturing;

(d) A continued high rate of growth of consumption of natural gas, particularly in manufacturing.

The report containing the projections made by the Department of National Development and Energy does not provide detailed results of the sensitivity of the figures to variations in the assumptions made. However, because of the strong relationship assumed between energy and GDP levels, and the relatively small price elasticity used, the projections are very sensitive to changes in GDP growth rates but comparatively insensitive to changes in relative prices. The heavy dependence of the estimates on the assumption of an average growth rate of real GDP of 3.5% per annum also needs to be borne in mind. In addition, it is now clear that the demand for electricity for aluminium smelting will not grow at the rate assumed by the Department. The reduced demand for aluminium products which has stemmed from the slow growth rates of OECD countries in recent years, combined with increases in electricity tariffs in some Australian states, has led to the cancellation of many of the previously announced plans for the construction of new smelters. As a consequence, it seems likely that less than half of the forecast demand for electricity for this purpose will now materialise in this decade. Deferment of the construction of additional potlines at existing smelters may also be expected if sluggish demand for aluminium products persists.

The sensitivity of demand for electricity in Australia to the demand for alumina, which is in turn heavily dependent on the growth of demand for aluminium products in OECD countries (Banks, 1979; Hojman; Owen and Fisher, 1982) illustrates the dependence of Australian resource development on growth rates in OECD countries. This is particularly true in the case of the United States and Japan, which are responsible for a high proportion of Australia's exports of mineral and agricultural products (McColl and Nicol, 1980). The recent slowing of OECD growth rates has led to a questioning of the forecasts of demand for other Australian mineral exports, including coal, natural gas and uranium. In addition, the prospect of slower OECD growth casts doubts on the projections of Australian energy demand based on an average annual real GDP growth of 3.5% per annum.

There are significant variations in the Government's latest projected rates of change in demand for various petroleum products. The average annual growth rate is high for LPG (9.6%), moderately high for aviation turbine fuel (3%), automotive diesel oil (2.7%) and aviation gasoline (1.9%), but the Department projected a decline in the use of motor spirit after the mid-1980s.

More recent forecasts made by oil companies in Australia imply a similar rate of growth of demand for petroleum products, but they also take account of the lower level of demand in recent years. According to these estimates, total demand will reach about 30 mtoe in the mid-1980s and 31 mtoe by the end of the decade (Australian Institute of Petroleum, 1981).

As demonstrated in the detailed workings of this paper, petroleum price increases averaging 2% per annum, combined with a price elasticity of −1, would cause demand for petroleum products to fall by the end of the 1980s. Demand would remain below recent levels at the end of the century even if a 4.5% GDP growth rate is attained. In the case of a 5% per annum rise in the price of petroleum products, a price elasticity of −1.0 would result in consumption falling by about one-fifth by the end of the 1980s and by between one-third and one-quarter by the end of the century.

BALANCE OF PAYMENTS EFFECTS

According to the Bureau of Mineral Resources, Australia's remaining recoverable (i.e. proved and possible) resources of crude oil, including condensate and liquid petroleum gas at 31 March 1981 exceeded 400 mtoe, a high proportion of crude oil being located in the Gippsland Basin (Bureau of Mineral Resources, Geology and Geophysics, 1981, Tables 1 and 2). Some time ago it was estimated that there was an 80% chance of discovering approximately a further 200 mtoe, a 50% chance of finding about twice that amount but only a 20% chance of finding about 700 mtoe (Esso, 1980). In recent years there has been a substantial increase in exploration activity without finding any new large sources of oil. However, sufficient quantities of crude have been discovered to maintain production at recent rates throughout the 1980s. When estimates of crude oil production are considered in relation to the revised estimates of demand, it appears that Australia will become no more dependent on oil imports until the 1990s.

For some time it has been expected that Australian crude oil production will decline in the 1990s, particularly in the later years of the century. Earlier expectations were that an otherwise increasing gap between consumption and production would be met by supplies from development of the oil shale deposits at Rundle, in Queensland (Esso, 1980). Although cost conditions were believed to be favourable at this site by world standards, the project was indefinitely deferred in 1981. This action appears to rule out the possibility of contributions to Australian oil supplies from shale sources for a considerable time.

A large amount of research is being carried out into the possibility of converting both black and brown coal to liquid fuels. It has been estimated that a high proportion of these resources is suitable for this purpose. However, it is now unlikely that commercial production of synthetic fuels from coal will take place for many years (Kirk, 1982). According to estimates made by the National Energy Advisory Committee, costs of fuels produced from coal liquefaction processes are likely to be higher than the costs of using shale oil (National Energy Advisory Committee, 1980a). It is therefore not surprising that there has recently been a decline in the interest being taken by foreign participants in coal liquefaction programs in Australia.

Early projections of rapidly increasing demands for imported oil, combined with increasing crude oil prices, caused many commentators to predict balance of payments difficulties for Australia later in this decade and beyond. For this reason, some arguments were advanced that the production of alternatives should be subsidised by governments. However, an analysis of the effects of a depreciation of the Australian currency which might be required if the Australian oil import bill should place a heavy burden on the balance of payments showed that it would be unlikely to be large enough to cause unmanageable economic problems (Folie and Ulph, 1979, Chapter 8).

Realisation that the foreign exchange which Australia seems likely to earn as a result of increased exports of coal, LNG, uranium and aluminium, all stimulated by the higher oil prices, led to a belief that the net effect of the changed international energy situation would be to cause upward pressure on the exchange rate for the Australian dollar. More recently, it has been shown that even if no new oil discoveries are made, Australia's oil import bill by the end of the century is unlikely to present unmanageable problems in the light of general prospects of increased Australian exports to the rest of the world (National Energy Advisory Committee, 1980b, Chapter 7).

PROSPECTS FOR COAL EXPORTS

The World Coal Study (Wilson, 1980) was carried out after the increase in oil prices in 1979. Based on the assumption that OPEC will not substantially increase its oil exports after the mid-1980s, the study forecast that the demand for steaming coal will rise rapidly. The study forecast a rapid increase in international coal trade. At the end of the century, its estimates of total demand for imports of coal ranged from the level of the IEA estimate to figures almost twice as high. The higher estimates arose from assuming greater limitations on oil availability and lower projections of installed nuclear electricity generating capacity.

According to the World Coal Study, Australia would play a leading role in meeting these demands, its projected exports in 2000 ranging from 140 to 300 million tonnes per annum. This range of estimates was based on preferences for sources of supply expressed by members of country teams which took part in the study. According to this pattern of preferences Australian coal exports in the year 2000 would no longer be dominated by sales to Japan. Western Europe would become much more important and Taiwan, South Korea and other Asian countries would together import more Australian coal than Japan (Wilson, 1980, Table 2.6).

Projections by the Joint Coal Board reflect the estimates made in the World Coal Study, although they do not envisage Australian coal trade growing as fast as assumed in the study's 'maximum potential' estimates. The Board projected that total exports of Australian black coal will increase by between three and four-and-a-half times up to 1990 and rise by at least a further 50% in the 1990s. In these estimates steaming coal exports are expected to rise at least tenfold by the end of this decade and reach sixteen times their recent levels by the end of the century.

International shipping costs have risen along with crude oil prices in recent years. Such rises adversely affect the prospects for trade in coal because of its greater bulk when compared with oil, which results in a considerable disadvantage in terms of landed costs when long distances are involved. Although the use of coal-fired ships, a number of which have already been ordered, may reduce this disadvantage in the long run (Bendall, 1979), it will be many years before such ships make a large impact on the demand for coal.

Additional questions relate to the relationship between the rate of economic growth in OECD and less developed countries in the next two decades and the level of international trade in coal. The outcome will be particularly important for Australian coal exports to such countries as Taiwan and South Korea, which are heavily dependent on international trade for their economic growth.

The assumed growth in coal demand by these countries is subject to great uncertainty. No participants from these countries took part in the World Coal Study and the estimates were apparently made by Japanese participants in the study. They assumed a continuance of very high growth rates in the two countries, despite the slow growth of world trade in recent years and the increasing scepticism expressed in a number of recent studies about the ability of less developed countries, including middle income countries such as Taiwan and Korea, to sustain recent growth rates (Burki, 1981).

Doubts may also be warranted about the future of Australian coal exports to western Europe. These exports will be more exposed to competition from United States, European and South African producers than trade in the Pacific region. The future level of Polish and perhaps Russian exports will be important in determining the markets which will be available in western Europe. In the case of Japan, future Chinese production and exports represent a major question mark.

These considerations suggest that caution should be expressed about the realisation of demand for Australian coal at the rates expressed in recent projections. This is particularly true as the projections extend further into the future, since the uncertainties associated with the assumptions made increase and the already wide range of plausible projections becomes even wider.

CONCLUSIONS

Despite the uncertainties associated with the growth of the international economy and the future availability and price of oil, the lower rate of development of Australia's mineral resources which is now expected to occur in this decade may still be expected to have a sizeable impact on the level of investment, output and exports, provided oil prices do not fall significantly and substantial real growth occurs in OECD countries.

Recent studies show that the imports required in both the investment and production stages of these developments will be small compared with the export income likely to be generated. Although income payable on foreign investment and repayment of capital borrowed from abroad will reduce the net foreign exchange receipts associated with the developments, there is likely to be pressure to appreciate the exchange rate for the Australian dollar later in the decade, unless the rate of inflation in Australia exceeds that in OECD countries generally.

The labour requirements associated with the developments, while small in relation to the total Australian workforce, may result in shortages in some skilled trades, particularly in the regions in which the developments are concentrated, such as the Bowen Basin in Queensland, the Hunter Valley in New South Wales and the Pilbara region in Western Australia.

Current and projected developments have already put considerable strain on the socio-economic environment in these areas. These issues are likely to continue to present major problems for the companies concerned with development, as well as for state and local governments.

Another issue, that of providing the finance required for the developments, including infrastructure, has also attracted much attention, particularly as the Commonwealth Government requires at least 50% domestic participation in most new mineral developments. While the funds required for ventures which do proceed are significant, they do not seem likely to present insurmountable problems in view of Australia's resource endowment and the apparent capacity of the companies concerned to obtain funds in international capital markets (see, for example, Stammer, 1982).

Given the distribution of sovereignty between the Commonwealth and state governments, it is perhaps not surprising that Australia was without a 'national energy policy' prior to 1973–74 or that it took a considerable period of time for any concerted movement towards such a policy to occur. This apparent inertia reflected a number of other circumstances, including Australia's relatively rich endowment of mineral resources and the availability of significant quantities of domestic oil, principally from Bass Strait.

There have been some important movements in the direction of a national energy policy, particularly the Commonwealth Government's decision to adopt import parity pricing for petroleum products in 1978. The pricing of other forms of energy, including coal, natural gas and electricity, are not based on sound economic principles, particularly in respect of capital inputs. While the changed energy situation has spawned a considerable debate on a number of important issues (Govett and Govett, 1980; Harris, 1981; Gruen and Hillman, 1981) there remain significant inconsistencies in the policies adopted by Commonwealth and state governments (Lloyd, 1981).

REFERENCES

Australian Institute of Petroleum, 'Forecasts of Demand for Oil Products, *Petroleum Gazette*, pp. 188-189 (December 1981).

Banks, F. E., *Bauxite and Aluminium: An Introduction to the Economics of Non-fuel Minerals*, Lexington Books, Lexington, Mass, USA, 1979.

Bendall, H., 'Coal-fired Turbines Versus Diesel: An Australian Context', *Maritime Policy and Management*, pp. 209-215 (1979).

Bureau of Mineral Resources, Geology and Geophysics, *The Petroleum Newsletter*, No. 84 (1981).

Burki, S. J., 'The Prospects for the Developing World: A Review of Recent Forecasts', *Finance and Development*, 18 (No. 1), (March 1981).

Esso Australia Ltd, *Australian Energy Outlook*, Sydney, 1980.

Folie, M. and Ulph, A., *Self-sufficiency in Oil: An Economic Perspective on Possible Australian Policies*, University of New South Wales, Centre for Applied Economic Research, C.A.E.R. Paper No. 6, 1979.

Govett, G. J. S. and Govett, N. H. (eds), 'The Australian Minerals Industry', *Resources Policy* (June 1980).

Gruen, F. H. and Hillman, A. L., 'A Review of Issues Pertinent to Liquid Fuel Policy', *Economic Record*, pp. 111-127 (1981).

Harris, S., *Social Aspects of Energy in Australia: A Social Science Literature and Research Review*, Seminar on The Challenge of Social Adjustment Posed by the Changing Position of Liquid Fuels, Joint Australian Academies, Canberra, 1981.

Hojman, D. E., 'An Econometric Model of the International Bauxite-Aluminium Economy', *Resources Policy*, pp. 87-102. (June 1981).

Kirk, J. F., 'The Prospects for Synthetic Fuels from Coal in Australia', Conference on Coal in Australia, Surfers Paradise, Queensland, 1982.

Lloyd, P. J., 'How Do We Make a Resource Policy' in *Resources in Australian Foreign Relations*, The Australian Institute of International Affairs, Canberra, 1981.

McColl, G. D. and Nicol, J. R., 'An Analysis of Australian Exports to its Major Trading Partners: the mid 1960s to the late 1970s', *Economic Record*, pp. 145-157 (June 1980).

National Energy Advisory Committee, *Strategies for Greater Utilisation of Australian Coal*, Australian Government Publishing Service, Canberra, 1980a.

National Energy Advisory Committee, *Liquid Fuels: Longer Term Needs, Prospects and Issues, 1980*, Australian Government Publishing Service, Canberra, 1980b.

Owen, A. D. and Fisher, L. A., 'An Economic Model of the U.S. Aluminium Market', *Resources Policy*, pp. 150-160 (June 1981).

Stammer, D., 'Resources Investments and the Capital Market', in J. W. Neville (ed.) *Resources Developments and the Australian Economy, Growth 32*, Committee for Economic Development of Australia, Melbourne, 1982.

Wilson, C. (ed.), *Coal—Bridge to the Future*, Ballinger, Boston, Mass., 1980.

PART XIII

The West European Supply Uncertainty

Chapter 19

Why Europe needs Siberian Gas

*Ferdinand E. Banks**

As things stand at present, the Soviets are scheduled to provide Western Europe with an additional 40 billion cubic meters (Gcm), or 40 × 35 (cubic feet/cubic meter) = 1.4 trillion cubic feet (Tcf), of natural gas per year for at least 20 years. Present deliveries are about 27 billion cubic meters, and from these the Soviets have gained about 3 billion dollars per year—which indicates an average price of about 3.15 dollars per 1000 cubic feet.

The probable cost of the new gas will be 4.60 dollars/Mbtu, cif the Czechoslovakian frontier, and the approximate transport cost from that point to Paris will be about 0.30 cents. On the other hand West Germany will pay an average price of about 4.75 dollars/Mbtu for delivered gas. Of the projected 40 Gcm, West Germany is to take 10.5 Gcm, France and Italy 8 Gcm each, the Netherlands and Belgium about 5 Gcm, Austria 3 Gcm, and Switzerland 1 Gcm. The opinion here is that once the gas starts flowing, these amounts will be adjusted upward, especially after the Dutch fields start declining at a rapid rate.

*The University of Uppsala, Sweden.

†This extract is taken from a long and detailed paper which examines the wider background to the issue discussed here and gives a comprehensive explanation of the assumptions used.

Next, let us get some idea of what this price means in terms of the price of oil. Thermally, or in terms of heating values, 1000 cubic feet of natural gas (= 1 Mbtu) is the equivalent of 0.178 barrels of crude oil. Thus it would take 1000/0.178 = 5620 cubic feet (= 5,620,000 btu) to get the same energy content as 1 barrel of oil. At a price of 4.75 dollars/Mbtu, this means that Soviet natural gas is selling at an oil equivalent price of almost 27 dollars per barrel. Here it can be noted that the price of Soviet gas will be allowed to fluctuate in phase with oil price changes, and the index used will apparently be one in which the weights are light fuel (heating) oil, 40%; heavy fuel oil 40%; and various crude oils, 20%. There is also supposed to be a floor price. By way of comparison, France recently agreed to pay Algeria an fob price of 5.1 dollars/Mbtu for the supply of 9.1 Gcm/year of liquified natural gas (LNG), and the cost of transport will raise the delivered price of these supplies almost to 6 dollars/Mbtu. Similarly, the cost of North Sea gas from the Statfjord field delivered to Emden (West Germany) is to be 5.5 dollars/Mbtu.

The indexing of the price of Soviet gas should be of considerable interest to academic economists with an interest in price theory, because it is probably the case that the larger the amount of gas purchased, the greater the downward pressure on the oil price, which in turn would reflect on the gas price. The total amount of gas that will be taken from the Soviets will replace about 1 million barrels of oil a day by 1987–88, and although this may not sound like a great deal today, it *could* be an important amount should a resurgence in world economic growth lead to excess demand on world oil markets. In addition, revenue from the present gas transaction could lead to a larger availability of Soviet energy supplies in the future, since they may find themselves with the financial resources to expand their investments in gas, oil, and coal. A major expansion of, for example, the Soviet coal sector will also result in large imports of steel and machinery from the OECD countries.

There are some other economic gains that have been reaped by Western Europe in association with the purchase of Soviet gas. At one time the Soviets could have borrowed a large part of the money to build this pipeline at an interest rate of 9.75% (= 7.75% in interest charges plus a 2% surcharge on the price of equipment delivered); but as things worked out they delayed making this arrangement, and now they will have to pay more than 11%, with the loan running for 10 years. These loans, incidentally, are to be repaid from gas revenues, and thus can be regarded as being of a very high quality—in contrast to many of the loans made by Western European banks to the Third World and Eastern Europe. In addition, an increased Soviet cash flow may decrease to some extent the possibility of a major Eastern Europe default.

More important is the present and future boost that will be provided Western European industry by Soviet orders for such things as pipe and machinery. These orders are to be divided among the various countries buying the gas on the basis of the amount of gas contracted for. At present total spending on the project comes to about 15 billion dollars: 5 billion for the pipe; 5 billion for compressor stations and communications facilities; and 5 billion to be spent on constructing the line, which will be done by the Russians themselves. But considering the nature of the enterprise, these amounts could easily escalate. The pipeline will cross permafrost, mountains, dense forests, and some 700 rivers. It will run for 5000 kilometers, starting at Urengoj (which is the largest gas producing complex in the world, with estimated reserves of 6.2 trillion cubic meters). Later on though, some gas for Western Europe may be taken from the even more northerly Yamburg field, which was originally scheduled to be the source of the new gas. The estimated reserves found in this field come to about 5 trillion cubic meters.

Some quantitative aspects of Western European energy dependence can now be discussed. Assuming that the transaction goes through as planned, West Germany's total energy dependence on the Soviet Union would increase from 3% to slightly under 6%. But this is still less than one-half of West Germany's dependence on Saudi Arabian oil, and there is no evidence anywhere that the USSR is an unreliable supplier of energy materials.

Since the first gas delivery agreements were signed by the Soviet Union with Austria, the Soviets have delivered a total of 130 billion cubic meters of natural gas to Western Europe. These deliveries have not been free of problems, because in especially cold winters there have been some pipeline freeze-ups and a fall in deliveries, but these have invariably been made up later. Conversely, proposed German deals with Iran and Algeria have fallen through at a very late stage of the negotiations.

Other OECD countries have also experienced difficulties in attempting to buy gas from non-European sources. As far as I can tell, the trouble is caused by some of the gas exporting countries insisting that gas prices should be close to or on a par with the price of oil; however the present supply and demand conditions on this market leave some doubt as to whether this position is justified. There is more than enough gas in the world to satisfy present consumption, and the people selling this gas are generally making large profits.

Table 19.1 provides some information about Western European energy dependence on Soviet natural gas. In connection with this table it should be mentioned that natural gas now supplies about 17% of Western Europe's energy consumption.

Table 19.1. The Gas Dependency of Some Western European Countries

	1980 (Percent)	1986 (Percent)
Western Europe	9	25
Austria	50-55	70
France	7	30
West Germany	16	30
Italy[a]	15-20	30
Belgium[a]	–	(5 Gcm)
Holland[a]	–	(5 Gcm)
Switzerland	–	20

Note: For Belgium and Holland the figures represent gas imports in billions of cubic meters.

[a] Estimated.

The last item to be taken up in this section is the actual construction of the pipeline, since the United States government has threatened to use the Strategic Exports Act to completely ban the use of United States technology on any aspect of the construction or operation of the line. What this mostly involves is an embargo on the moving parts (e.g. rotor shafts and blades) of the gas turbines that would be employed in the 40 or so compressor stations that will be built at intervals of 100-120 kilometers along the pipeline. European firms such as John Brown of Edinburgh, AEG Kanis of West Germany, and Nuovo Pignone of Italy import these parts from General Electric in the United States.

The net result of this kind of policy is likely to be a slowing down of the building of these lines—at best. It is probably true that European companies can be prevented from selling United States technology in any form to the Soviets, but they can hardly be prevented from helping the Soviets construct factories to produce the embargoed equipment, assuming that such help is required. As I make clear in *The Political Economy of Oil* (Chapter 8), the Soviets have shown time and again that in the matter of complicated military hardware they can produce equipment on a level with any available in the world; and although it is not fully appreciated, the Soviet Union manufactures at least two-thirds of the material needed to support its production of oil and gas. In fact the Soviets have already announced that a factory in Leningrad has started producing gear capable of replacing that which cannot be obtained from General Electric. It is also true that, among others, the French manufacturer Alsthom-Atlantique would be capable of constructing the rotor blades and shafts needed by the Soviet compressor stations at fairly short notice, although eventually this might cause some legal and political problems.

* * *

We can now look at some, but not all, alternatives to Soviet natural gas. The pipeline from North Africa to Italy has already been mentioned, and obviously this line—whose terminus will be Bologna—could be duplicated. A pipeline between Algeria and France with a terminus in the vicinity of Marseilles has been discussed for many years; and France has recently signed an agreement to receive 9.1 Gcm/year of LNG from Algeria, for which an fob price of 5.1 dollars/Mbtu will be paid. Shipping and regasification will probably add about a dollar to this amount, and the gas contract will include an indexation formula relating the gas price to a basket of light fuel oils.

The interesting thing here is not just the price that will be paid for this gas, which is higher than the price of Soviet gas, but the fact that Algeria has distinguished itself over the past few years by insisting that purchasers of Algerian gas must pay full parity prices—that is, prices that are equivalent to the oil price. This attitude has caused some bad feeling between Algeria and its potential customers, in particular France and the United States. There is also the basic reality that LNG systems are generally considered to be economically inferior to pipelines when the length of the pipeline is 8000 kilometers or less—although there are arguments that the crossover from pipeline to LNG could be reduced to 5000-6000 kilometers.

CONCLUSIONS

This paper has reviewed some economic and, to a limited extent, political problems associated with the proposed large increase in Soviet natural gas supplies to Western Europe. My conclusion is that the more of this gas purchased the better. What I have not pointed out is that once the tap is opened on Soviet gas, and it shows signs of staying open, it may be possible to make use of pipelines through southern Russia to market the huge gas reserves of Iran. Also, since history makes it amply clear that the appetite for gas grows with its consumption, a way might be found to stop the scandalous waste of Gulf gas reserves (by 'flaring'); and perhaps through liquification make these available on a larger scale to Europe and elsewhere. If so, the world energy picture would certainly assume a saner composition—not the least from an environmental point of view.

Chapter 20

An Opportunity for Gulf Gas

*Melvin A. Conant**

If a very large energy source, such as natural gas from the Middle East, by long-distance pipeline were to suddenly become available what would be the consequences? Limitations on gas use would become subject to total revision. That is the prospect, that is the challenge to realize upon a potential source of such size as to make earlier estimates of global gas consumption irrelevant.

The gas reserves of the Gulf region, known only for its oil, are vast; conservative estimates place Middle East proven reserves at 18,140 BCM with 1980 production at 50 BCM. To compare, the largest single source of proven gas reserves is currently the USSR, which lists proven reserves at 37,000 BCM, with a 1980 production of 435 BCM.

Since the Middle East can absorb only a fraction of what could in time be made available, the remainder must either be exported, flared, or kept unexploited to the enormous, eventual disadvantage of both the owner and the potential market.

*President, M. Conant and Associates, Washington.

†Anyone interested in international gas trade has to acknowledge the seminal contributions to the subject by Jonathan P. Stern.

Of the great gas markets of the world: North America, the Soviet Union and Europe, and Japan, only Japan and Europe will be significantly dependent on external sources for the larger share of their supply. The Pacific Basin suppliers mainly will meet Japanese needs; the Middle East is by far the most prolific gas source for Europe.

Western Europe as the market for large gas flows from the Middle East has been an ignored option; economic, political and security considerations have seemed too problematical, and the market estimates too limited.

Western Europe's gas consumption could rise from 220 BCM in 1980 to 310–384 BCM by the year 2000 in light of anticipated general increases in Europe's energy demand. In 1980 Europe imported 34 percent of its gas consumption; France 38 percent, and Italy 28 percent, from sources outside Europe. Imports necessary by 1990, from outside Europe could reach 48 percent for West Germany, 83 percent for France and 68 percent for Italy. By the year 2000, with moderate economic growth, Total European dependence on gas imports could reach 50 percent.

The market opportunity is evident. Who will, therefore, be the suppliers? Norway, the USSR and Algeria are projected by the European Community to remain Europe's leading suppliers in 1990. Are these three countries Europe's only, or preferred, options?

The current absence of a large alternative supplier makes almost certain the prospect that the Soviet Union will be the pre-eminent international gas supplier to Europe without a close competitor. The USSR already supplies almost 12 percent of total consumption. Current and prospective supply contracts indicate that as much as 100 BCM of Soviet gas may flow to Europe by the year 2000. This could approximate one-fourth of total gas consumption or nearly equal to all the supply now anticipated fron Norway, Algeria, Libya, Nigeria and even the East Arctic islands of Canada.

From the strategic perspective of the Europeans, it is necessary to diversify gas supplies as broadly as practicable. There are apprehensions within the European Community but even more in the US about the prudence of Europe becoming excessively dependent on the USSR. Problems with Algeria over price and dependable supply may have made future expansion of Algerian-European gas trade an unattractive option.

Another, even larger, source of gas from the Middle East would offer Europe a diversified source to Soviet supply. While troubles in the Middle East make comparisons with Soviet supply difficult to make, there are, however, arguments which, over time, could become compelling for Middle East supply. A central consideration is the mutual interest of Europeans and of Middle Easterners to expand upon and envelop their energy needs as a core to a larger and more consequential economic, financial and political relationship. If this can occur around natural gas flows then there will be a compelling case for asserting that dependable supply has a concurrent meaning for all parties. The recently formed Gulf Cooperation Council (GCC) might facilitate negotiations toward that end.

Two strategic corollaries need to be kept in mind. *Firstly*, LNG can in no way be the logistics vehicle for transporting large flows of gas. Piped gas will be required largely because of its generally lesser expense, its greater capacity and ease of expansion, and environmental safety. With the exception of supplies from the USSR (with the possibility of renewed consideration of an Iranian connection via the USSR) the trans-Mediterranean (Algeria to Europe) pipeline (where the partners have had greatest difficulty in agreeing on a price), the Bechtel/Nigeria proposal of piping gas to southern Europe, and the intended Segamo pipeline (Algeria to Spain), *all gas supply from the Middle East/North Africa to Europe is presently limited to consideration of the LNG option.* Yet, even the most optimistic forecasts of LNG to Europe make clear the greater costs inherent in LNG processing and shipment over ordinary pipeline supply. *LNG is now and will likely remain a very low priority form of energy supply when the consumers and producers have a feasible pipeline option.*

The *second* strategic corollary is that there are two immense reserves of natural gas in the region—the Iranian and what we can describe as 'Gulf' or Arab gas. Either source would be more than sufficient to initiate the energy revolution for Europe. There are ramifications and consequences to all parties if one or the other is chosen.

In earlier years Teheran had undertaken to begin pipeline supply to Europe via the USSR—a commitment rejected to date by the present government. It could be renewed thus pre-empting gas exporters of the Arab Gulf, were Europe to have no other option. Moreover, there is now known to be a renewal of Turkish-Iranian undertakings to examine the feasibility of Iranian gas piped to Europe transitting Turkey.

The strategic implications of the choice between Iranian versus Arab supply are embedded in what one thinks the course of revolutionary Iran is likely to be, and the conclusions one draws about the prospects for stability on the Peninsula and in the Gulf: would piped supply to the Mediterranean be less risky than through Turkey? The options need to be examined.

If the Arabs wish to be the prime supplier, they must move first. Current Middle East export commitments (LNG) for 1990 are only

3.0 BCM from Abu Dhabi and the same for 2000. Potential exports of Middle Eastern gas could reach 58.5 BCM by 1990 from Abu Dhabi, Qatar, and Iran, and perhaps 375 BCM by 2000 from the same three plus Saudi Arabia. These amounts would appear insufficient were they compared to the level of gas which could be supplied to Europe through pipeline.

The high cost of the logistics system, including not only the technology, design and construction but the anticipated cost of capital, will be a large economic issue of the gas revolution. Most consequential, however, will be the price the producer will ask for its supply (a diminishing resource) and the size of the potential market. Each is closely related to the other; each is dependent on the other.

If, however, suppliers accept, as a basic premise, that the price of gas imported into Europe has to be competitive with alternative fuels at the burner-tip, then the potential of the European market becomes almost immeasurable. Such a concession, admittedly, means a lower return to the producer than he has become accustomed to receiving from both LNG and oil. With the exception of the unlikely option of leaving gas in the ground, does the producer have an alternative? Particularly, as many face the prospect of declining oil reserves, and therefore revenues?

Producers and consumers must be convinced of the dependability of natural gas supply across national boundaries. Certainly the Iraqi experience, both before and during its current war with Iran, in trying to maintain exports through pipelines will not make acceptance easy of the dependability of pipeline gas supply. Contractual arrangements can, however, be developed which provide maximum incentives for all parties—but never complete assurance—to maintain flow, as evident in the discussions of the ramifications of Iranian supply via the USSR, of trans-Mediterranean supply via Tunisia, and of European supply obtained from the USSR, transitting East Europe; or of the still possible transitting of Alaskan gas through Canada by pipeline. As mentioned earlier, were Europe's need for natural gas from the Middle East to promote an evolution of complementary economic, financial, and political relationships, the overall dependability of European energy supplies, both oil and gas, from the region could be increased.

It is impossible to scale the potential market at this time, in the absence of market studies whose assumptions are perhaps unprecedented: virtually unlimited supplies of natural gas at a competitive price. The 'Groningen' syndrome is worth keeping in mind. Prior to its exploitation, which energy analysts predicted the quick and large role accorded the discovery in the economy of the Netherlands (and of bordering countries)? We are discussing more than a hundred 'Groningens'.

The full development of Gulf gas—assuming the terms for gas supply are competitive—will be the single most important consideration in world energy before the end of this century.

PART XIV

New Strategies in OPEC Planning

Abstracts (see Section 5)

Chapter 21

The Impact of Hydrocarbon Processing in OPEC Countries†

*Fereidun Fesharaki and David T. Isaak**

If refinery expansions in OPEC and Gulf nations proceed as scheduled, if Egypt and Mexico complete their expansions, and if OECD plants under construction are completed, world refining capacity will expand by about 6 million b/cd; if other developing countries complete their plans, expansions could total 7.4 million b/cd. If total world demand for oil products remains static, this would imply a drop in the world average capacity utilization rates to about 68–69 percent. This could lead to refining losses even greater than those seen in recent years. Massive refinery closures seem likely in the wake of such losses. Before we examine the likely candidates for the scrapyard, however, a closer examination of the capabilities and economics of OPEC and Gulf refineries is in order.

*Dr Fereidun Fesharaki is Coordinator, and David T. Isaak, Assistant Coordinator of the *OPEC Downstream Project*, Resource Systems Institute, East-West Center, Honolulu, Hawaii 96848, USA.

†This detailed paper draws heavily on the conclusions of the authors' new book, *OPEC, the Gulf and the World Petroleum Market: A Study in Government Policy and Downstream Operations*, Westview Press, Colorado (forthcoming, 1982).

Table 21.1 Projected Product and Crude Exports from OPEC in 1986 (Thousand b/d; rounded to nearest 10,000 b/d)

Country	*(1) Crude[a] Production*	*(2) Refinery[b] Runs*	*(3) Product[c] Consumption*	*(4) Crude Exports (1–2)*	*(5) Product Exports*	*(6) Total Exports (4+5)*
Algeria	900	710	120	190	590	780
Ecuador	200	180	110	20	70	90
Gabon	200	20	40	180	(20)	160
Indonesia	1,640	850	510	790	340	1,130
Iran	3,170	1,120	610	2,500	510	2,560
Iraq	3,090	350	390	2,740	(55)	2,685
Kuwait	1,500	700	70	800	630	1,430
Libya	1,500	330	130	1,170	200	1,370
Nigeria	2,060	240	210	1,820	30	1,850
Qatar	500	60	10	440	50	490
Saudi Arabia	6,960	1,740	640	5,220	1,100	6,320
UAE	1,800	320	120	1,480	200	1,680
Venezuela	1,840	1,360	410	480	950	1,430
Neutral Zone	500	30	0	470	30	500
OPEC	25,860	8,010	3,370	17,850	4,625	22,475

[a]By implicit growth rates from 1985 and 1990 projections in Fesharaki, F. and D. T. Isaak, *OPEC, The Gulf, and the World Petroleum Market*, Westview Press, Boulder, 1982. (Forthcoming.)

[b]Taken as 91 percent of calendar day (85 percent of stream day) capacities in Table 21.8 of above volume.

[c]By implicit 1985–1990 growth rates from projections in Table 21.4 of above volume.

FUTURE OPEC PRODUCT EXPORTS AND GULF REFINERY FLEXIBILITY

Given production levels and oil product demand in OPEC countries, it is possible to assess likely levels of product exports from OPEC nations around 1986. This exercise is performed in Table 21.1. The capacity utilization assumed is high (85 percent of design capacity, or 91 percent of calendar day capacity), but not unreasonably so, and is well below the physical capabilities of the refineries.

The problem of assessing the possible product mixes from the refineries is more complex. The output mix is affected by the kinds of crude processed, the size and types of processing units, and the operating strategy employed. The finer points of product blending specifications can produce constraints that are apparent only to the quality control engineer. An accurate assessment of refinery capabilities would require a very large simulation model for each refinery.

For our purposes here, a simpler approach is needed. To determine the possible output mixes, we have employed *PRYMO*, a Petroleum Refinery Yields Model under development at the East-West Center. *PRYMO* is not a linear programming model, but rather a physical simulation model based on the correlation chart type of analysis introduced by Gary and Handwerk. Unlike a linear programming model, *PRYMO* models what a refinery can do, not what it ought to do.

PRYMO can be run in four modes: Operator control, light product maximizing, middle distillate maximizing, and heavy fuels maximizing. In the last three modes, the model uses crude oil characteristics curves and the refinery configuration to try to maximize the product group in question. At present, the model can simulate atmospheric distillation, vacuum distillation, catalytic cracking, visbreaking, coking, distillate hydrocracking, and residuum hydrocracking, though with varying degrees of accuracy. In addition, it can correct for refinery energy use.

To clarify questions regarding future output capabilities, we have applied the model to all Gulf refineries, present and planned, in the configurations planned for 1986. We hope to expand the scope to cover other areas in future work.

Each of the refineries was simulated in light, middle, and heavy fuel maximizing modes, running the appropriate crude oil. Crude runs were taken at 91 percent of calendar day capacity. The refineries simulated, and their assumed throughputs, are shown in Table 21.2. The results of the simulations, aggregated by country, are shown in Table 21.3.

The product mixes shown in each mode represent an extreme; in

Table 21.2 Projected Operations of Gulf Refineries in 1986

Country	*Name/Location*	*Estimated Capacity b/cd*	*Estimated Crude Runs b/cd*
Iran	Abadan	563,000	515,000
	Esfahan	219,000	200,000
	Kermanshah	19,500	18,000
	Masjid-e-Sulaiman	73,000	66,000
	Shiraz	44,000	40,000
	Tabriz	87,000	80,000
	Tehran	229,700	209,000
Iraq	Baiji	39,500	127,500
	Basrah	141,700	129,500
	Daura	78,000	71,000
	Haditha	7,600	7,000
	Kirkuk	2,200	2,000
	Khanaquin	13,000	12,000
	Mufthia	4,900	4,500
	Qaiyarah	2,200	2,200
Kuwait	Mina Abdulla	311,400	283,000
	Mina Al-Ahmadi	250,000	227,000
	Shuaiba	205,000	187,000
Qatar	Umm Said I	13,700	12,500
	Umm Said II	46,500	42,500
Saudi Arabia	Jeddah	98,000	90,000
	Juaymah	233,000	213,000
	Jubail	233,000	213,000
	Rabigh	302,000	276,000
	Ras Al Khafji/Mina Saud	80,000	32,000
	Ras Tanura	507,000	462,000
	Riyadh	112,000	103,000
	Yanbu (Petromin-Mobil)	274,000	250,000
	Yanbu (Petromin)	158,000	144,500
UAE	Umm Al Nar	70,000	63,800
	Ruwais I & II	284,000	259,000
Bahrain	Awali (BAPCO)	274,000	250,000
Oman	Muscat	46,500	42,500

practice, most refineries will probably not be operated in one of these modes, but rather somewhere in between. The actual operating strategy will be set by domestic product requirements and product prices on the export market.

What the model shows is that the Gulf, taken as a whole, will have a fairly flexible refining system by the late 1980s, capable of substantial adjustments to meet changing demand patterns. The amount of flexibility varies considerably between countries, however, both as a result of differences in the sophistication of the refineries and as a result of differences in the characteristics of crude oils. Additional cracking facilities planned for Kuwait may enhance the Gulf's flexibility even further. This level of flexibility in the Gulf refining system allays to some degree earlier fears that the Gulf refineries might be dumping large volumes of fuel oil on the market in the late 1980s.

Using our simulation model and other information available to us we have estimated the mix of product exports from the Gulf in Table 21.4. The table shows two scenarios of product exports in 1986. Both scenarios assume that the refineries of Iran, Iraq and the UAE will run in the middle distillate mode to attempt to meet domestic demand. Other refineries are assumed to run in light products mode in Scenario I and middle distillate mode in Scenario II.

The refinery flexibilities shown in Table 21.4 are self-explanatory. It indicates that the Gulf refineries are likely to have a range of 0.7 to 1.1 mmb/d of light product exports; 0.4 to 1.0 mmb/d of middle distillate exports and between 0.9 and 1.0 mmb/d of heavy product exports by 1986. The real significance of the exercise is to show that the Gulf exporters are able to watch the demand developments and marketing prospects in order to devise their refining production strategies. (Note: product export data in Table 21.1 may not exactly match data in Table 21.4. The difference is that one considers refining runs and the other refining output.)

ECONOMICS OF GULF EXPORT REFINING

We are often asked by oil analysts why the Gulf nations are moving into export refining when it is 'clearly uneconomic'. Our answer is that such an assertion is far from clear. National oil companies and private oil companies have different goals and different investment options. Furthermore, many Gulf export refineries will also be serving domestic markets to varying degrees; when a government both pays for refinery construction and then subsidizes domestic product consumption by holding prices below world levels, common business criterion of what is economic hardly applies. A 'return on investment' type of approach may be appropriate when comparing two investments available to a single investor, or even when compar-

Table 21.3. Projected Flexibility of Gulf Refineries in 1986 (Barrels per day)

	Heavy Mode		*Middle Mode*		*Light Mode*	
	Volume	*%*	*Volume*	*%*	*Volume*	*%*
Iran						
Light	246,163	22.3	232,235	21.0	432,271	38.6
Middle	247,033	22.3	494,458	44.8	232,366	20.8
Heavy	613,170	55.4	377,950	34.2	455,012	40.6
TOTAL	1,106,366	100.0	1,104,644	100.0	1,119,649	100.0
Iraq						
Light	86,816	25.6	78,366	23.1	157,290	45.6
Middle	82,074	35.2	161,371	47.6	72,659	21.1
Heavy	170,248	50.2	99,482	29.3	114,755	33.3
TOTAL	339,138	100.0	339,219	100.0	344,715	100.0
Kuwait[a]						
Light	111,724	16.5	105.683	15.4	221,349	31.9
Middle	124,102	18.3	271,987	39.8	122,737	17.7
Heavy	442,768	65.2	306,871	44.8	350,110	50.4
TOTAL	678,595	100.0	684,482	100.0	694,196	100.0
Qatar						
Light	14,983	29,5	12,895	25.3	19,740	39.0
Middle	14,156	27,8	23,188	45,5	12,364	24.4
Heavy	21,677	42.7	14,882	29.2	18,543	36.6
TOTAL	50,816	100.0	50,965	100.0	50,647	100.0
Saudi Arabia[b]						
Light	352,639	21.1	460,559	26.8	637,660	37.0
Middle	383,379	22.9	662,013	38.4	360,372	21.0
Heavy	935,739	56.0	598,457	34.8	721,436	42.0
TOTAL	1,671,757	100.0	1,721,029	100.0	1,719,468	100.0

ing similar investments to similar investors, but it is questionable practice when comparing governments and firms.

There is little question that the Gulf governments will subsidize the construction of their refining industry; the United States and some European governments have provided various subsidies to refiners at various points. What we hope to assess here in this detailed hypothetical study is whether the prevailing subsidies are sufficient to make Gulf refineries competitive on the world market.

Table 21.3 *(continued)*

	Heavy Mode		Middle Mode		Light Mode	
	Volume	*%*	*Volume*	*%*	*Volume*	*%*
UAE						
Light	85,135	27.4	84,059	27.1	180,589	56.5
Middle	83,934	27.0	184,677	59.7	74,497	23.3
Heavy	141,380	45.6	40,694	13.2	64,473	20.2
TOTAL	310,449	100.0	309,430	100.0	319,559	100.0
Bahrain						
Light	50,363	21.3	79,286	32.5	172,321	67.9
Middle	54,753	23.1	151,355	62.1	52,283	20.6
Heavy	131,562	55.6	13,272	5.4	29,219	11.5
TOTAL	236,678	100.0	243,913	100.0	253,823	100.0
Oman						
Light	7,896	20.3	6,673	17.2	10,848	28.0
Middle	9,464	24.3	16,107	41.6	8,720	22.5
Heavy	21,593	55.4	15,976	41.2	19,184	49.5
TOTAL	38,953	100.0	38,757	100.0	38,751	100.0
Total Gulf						
Light	955,719	21.6	1,059,706	23.6	1,832,068	40.3
Middle	998,893	22.5	1,965,147	43.7	935,998	20.6
Heavy	2,478,137	55.9	1,467,584	32.7	1,772,743	39.1
TOTAL	4,432,749	100.0	4,492,437	100.0	4,540,809	100.0

Note: Does not include output of NGL facilities.

[a]Without addition of cracking facilities at Mina-Al-Ahmadi and Mina Abdulla.

[b]Since configurations of the new refineries at Jubail and Juaymah are unknown, the capabilities of these two plants have been assessed as if they were equivalent to the Petromin-Mobil refinery at Yanbu.

CONCLUSIONS

The world petroleum market is continuing to undergo structural changes. One such structural change is in the refining industry, where the turmoil is inflicting great financial damage to the industry. The crisis in refining may not have caught the attention of the media or politicians, but nevertheless its impacts are serious and far reaching.

Table 21.4. Scenarios of Petroleum Product Exports from the Gulf, 1986 (000s Barrels Per Day)

Type of Product		Scenario I[a]			Scenario II[b]		
		Light	Middle	Heavy	Light	Middle	Heavy
Iran	Production	232.2	494.5	378.0	232.2	494.5	378.0
	Consumption	98.4	372.6	135.5	98.4	372.6	135.5
	Exports (Imports)	133.8	121.9	242.5	133.8	121.9	242.5
Iraq	Production	78.4	161.4	99.5	78.4	161.4	99.5
	Consumption	68.8	197.0	128.7	68.8	197.0	128.7
	Exports (Imports)	9.6	(35.6)	(29.1)	9.6	(35.6)	(29.1)
Kuwait	Production	221.3	122.7	350.1	105.7	272.0	306.9
	Consumption	36.1	30.1	5.6	36.1	30.1	5.6
	Exports (Imports)	185.2	92.6	344.5	69.6	241.9	301.3
Qatar	Production	19.7	12.4	18.5	12.9	23.2	14.9
	Consumption	4.3	5.5	—	4.3	5.5	—
	Exports (Imports)	15.4	6.9	18.5	8.6	17.7	14.9
Saudi Arabia	Production	637.7	360.4	721.4	460.6	662.0	598.5
	Consumption	109.0	227.7	301.8	109.0	227.7	301.8
	Exports (Imports)	528.7	132.7	419.6	351.6	434.3	296.7
UAE	Production	84.1	184.7	40.7	84.1	184.7	40.7
	Consumption	20.2	97.6	4.5	20.2	97.6	4.5
	Exports (Imports)	63.9	87.1	36.2	63.9	87.1	36.2
OPEC Gulf Exports		936.6	405.6	1032.2	637.1	867.3	862.5
Bahrain[c]	Production	172.3	52.3	29.2	79.3	151.4	13.3
	Consumption	3.2	3.1	0.7	3.2	3.1	0.7
	Exports (Imports)	169.1	49.2	28.5	76.1	148.3	12.6
Oman[c]	Production	10.8	8.7	19.2	6.6	16.1	16.0
	Consumption	6.4	8.3	0.1	6.4	8.3	0.1
	Exports (Imports)	4.4	0.4	19.1	0.2	7.8	15.9
Gulf–TOTAL EXPORTS		1110.1	455.2	1079.8	713.4	1023.4	891.0

Source: OPEC Downstream Project.

[a] Iran, Iraq, and UAE in Middle distillate maximizing mode; others maximizing light products.

[b] All refineries in Middle distillate mode.

[c] No demand forecasts were given for Bahrain or Oman. We have therefore estimated their consumption by growing aggregate 1979 consumption at 5 percent per annum, and assuming a constant demand mix. For Bahrain, the mix is determined from *OAPEC Statistical Bulletin, 1979*, OAPEC, Kuwait, 1980; for Oman, from *United Nations Energy Yearbook 1979*, United Nations Statistical Office, New York, 1980.

The problems of the world refining industry are likely to continue and worsen over the 1980s. Massive excess capacities are already with us and the impact of OPEC refineries will make a bad situation worse. OPEC refineries—whether we classify them as economic or not—are being built and will become a major force in the oil market. Current financing feedstock prices and contractual arrangements will ensure that OPEC product exports will be price competitive and can be marketed. The marketing of OPEC products will take place either through joint venture partners or independently. If and when the crude oil market tightens again, OPEC nations will be in a strong position to package crude exports with product exports.

The international refining industry will be affected differently around the world—depending on ownership, staying power, crude sources and export markets. But it is quite clear that something has to give; the present situation cannot go on for long. Unfortunately, the market induced correction mechanisms will not work smoothly. Many refiners will resist scrapping or closures, hoping for a miracle: the large upswing in demand for oil. Many new refineries will be built, particularly in the Third World, on the mistaken notion that their own refineries will enhance security of supply and provide value added.

The major oil companies have been quicker to respond to the over-capacities. Independents may want to fight it out, take losses for a few years and hope for the best. For all their bravery, our conclusions remain unchanged: massive scrapping must take place to increase capacity utilization rates from the current dismal levels to profitable levels. Unsophisticated refineries cannot hope to survive the highly competitive market of this decade. We realize that this is a painful process for many refinery owners, but the longer they stay in the market, the more losses they will make and they will yet have to scrap. Governments in some countries might be tempted to intervene to respond to political pressures for saving inefficient refineries. This possibility can lead to grim consequences as other governments will be forced to do the same to help their refineries, and the market could end up with even larger losses and dislocations. It is best, we feel, to accept the new realities and try to 'rationalize' the inefficient industries rather than prolong the agony.

Finally, we would like to point out that for OPEC nations, every barrel of oil refined is a barrel not available as crude exports. Thus, product exports will replace a portion of crude exports but not necessarily lead to higher aggregate exports. This means that attempts to use administrative/tariff barriers to stop the flow of OPEC products will not be successful and would lead to further delays in the readjustment process.

PART XV

Breaking International Deadlock in the Developing World

Abstracts (see Section 5)

Chapter 22

The Energy Crisis and the Third World: An Opportunity for the West

*Charles K. Ebinger and Harry Luzius**

The oil price increases since 1973 have had a devastating impact on the economies of the Third World nations. Whereas on the eve of the first OPEC shock, the oil import bill of the Third World nations was $3.5 billion, by 1981 it had escalated to around $35 billion. The less developed countries, which are sizeable net importers of food and petroleum-derived fertilizers, simultaneously saw the prices of these goods skyrocket. Oil-induced inflation in the industrialized world after 1973 sparked a major surge in the price of manufactured goods imported by the Third World. Furthermore, unable to off-set the cost of expensive oil imports, many states experienced a further erosion in their financial position through ever-increasing debt-service obligations. Because most international debt is denominated in dollars, the strength of the US currency during 1981–82 has made the financial burden even more severe. The fact that the oil-induced recession in the industrialized world led to decreases of purchases from LDCs further eroded LDC foreign exchange receipts, thus weakening their already precarious financial position. As a result of these events, the non-oil developing countries' balance of payments deficits mushroomed from $8 billion in 1972 to

*The Center for Strategic and International Studies, Washington, USA. With the assistance of James Hayes and Ellen Hall.

$80 billion in 1980 and about $100 billion in 1981. While private commercial lenders deserve credit for helping the LDCs deal with this difficult situation, their risk exposure in the LDCs alone rose from $32 billion in 1970 to $149 billion in 1978 and $284 billion in 1980.

A WIDE DIVERSITY

Although one frequently refers to the OIDCs (Oil Importing Developing Countries) as a cohesive group of nations, this is misleading owing to the wide diversity in their resource base, geography, demography and development objectives. Whereas some of these countries, such as Burundi, Haiti, and Upper Volta, are abysmally poor with per capita incomes below $200, Brazil and South Korea in contrast, have large manufacturing sectors and annual per capita incomes in excess of $1500. While many of the poorer nations of the Sahel have very low levels of energy consumption per capita, the upper tier OIDCs often have energy consumption levels 60–80 times as large.

Of the 120 Third World nations (including the 28 oil exporters), 92 import petroleum. Although many OIDCs import relatively small amounts of oil, 50 of them depend on imported oil for 90 percent of their commercial energy consumption, while most of the rest, with the exception of India, Pakistan, S. Korea and Zambia, depend on oil imports for between 50-90 percent of their commercial energy consumption.

Furthermore, it is critical to remember that while energy analysts concentrate on the impact that the oil crisis has had on the economies of the OIDCs, at least one-half the world's population continues to depend on wood and other traditional fuels, including crop and animal wastes, as their principal source of fuel. As the IBRD notes, in many OIDCs, industry also places primary reliance on fuelwood. Indeed in Mali, Tanzania, Nepal, Ethiopia and Haiti, traditional fuels account for over 90 percent of total energy consumption.

A CYCLE OF ENDEMIC DEBT

Total LDC debt to all lenders having risen from $67.7 billion in 1970 to in excess of $500 billion in 1981, combined with the fact that 80 of the largest US banks have single LDC exposures greater than 30 percent of their capital, raises profound implications for the stability of the international financial system if one or more major debtor nations were to default on their obligations. Even if they do not, rising debt-service ratios in excess of 50 percent in at least 13 countries threaten to burden future generations throughout the Third World with a cycle of endemic debt. In such an atmosphere, little social, economic and political development will occur, thereby making political chaos a distinct possibility.

NON-FOSSIL FUELS

The dramatic increases in the price of oil, combined with the inability of high cost electrification programs to reach many Third World citizens, have led to an increased reliance on non-fossil fuels (wood, bagasse, cotton stocks, etc.) with serious attendant direct and indirect damage to the global environment. The threat of massive deforestation is now a reality in many parts of the world. This is especially true in Nepal, parts of South Asia, large portions of Africa and in much of the Caribbean.

Shortages of non-commercial fuels are not a new problem in those portions of Latin America, Asia and Africa where large population growth and the need to clear land for agricultural production have long exerted pressure on forests. What has changed since 1973, however, is the fact that whereas previously the pattern of energy utilization in LDCs had been a shift from the use of firewood, dung or crop residues to conventional fuel sources, today the higher costs of conventional fuels relative to traditional ones has spurred an increased use in these resources for fuel needs. This development has negative economic and environmental effects. If more wood is taken from forests each year than the annual increment of forest growth, forested areas decrease. This leads to the burning of crop and animal dung which precludes these materials from being used as fertilizers. Soil fertility is reduced, desertification and deforestation accelerate, food production declines, export earnings are reduced and domestic inflation increases. This, in turn, pushes these nations further into a reliance upon fossil-fuel based fertilizers to make up for lost soil productivity. Greater dependence on such fertilizers implies greater imported supplies of these products. With the top soil barren, rains then lead to further siltation of rivers which over the years can ruin the use of large scale hydroelectric projects designed to generate energy and to help irrigate the land. Finally, deforestation reduces the earth's capacity to absorb the extra carbon dioxide caused by burning fuels, which can raise global temperatures thus affecting the weather, and patterns of crop production.

For the rural poor, the increasing scarcity of fuelwood resources intensifies the hardships of daily living. As forests vanish, time has to be devoted to the gathering of fuelwood, the burden of which generally falls on women and children. The attendant result is that each household or village has less people working on the land which further reduces agricultural productivity. On the basis of survey research conducted by the author in northern Pakistan, one or more family members have to devote 10-12 hours per day simply to gather enough debris for one night's cooking and/or heating. Still another consequence of the fuelwood crisis has been lower nutritional standards, as fuel shortages have reduced the number of cooked meals per day.

ENERGY SUBSIDIES

For the urban poor, fuel (mainly charcoal and Kerosene) account for a large part (often 35-50 percent) of total expenditure second only to food. For this reason, governments often subsidize fuel prices. Increases in energy prices involve great potential hardship, especially if they occur in concert with rising food prices. As a result, governments are reluctant to raise prices even when confronted by IMF or bilateral donor pressure. On a number of occasions when governments have responded to the conditionality of IMF credits and raised energy prices, riots have broken out and governments have even been toppled.

In observing the energy problem from both a macro- and micro-level, one can assert that the stake of the industrialized world in LDC development is not limited to the humanitarian concern for the alleviation of poverty and misery. Although development does not guarantee domestic political stability, sporadic or frustrated development is a certain producer of instability. As Deese and Nye assert, 'Even with prudent macro-economic management, governments may face untenable political opposition to the direct and indirect effects of increasing energy prices'. Unanticipated and sudden changes in governments can upset current geopolitical alliances. In a world already driven by ideological rivalries, the competition for vital energy supplies can jeopardize the prospects for international security.

It can thus be argued that energy sector planning is a necessity for developing countries, since energy costs or supply difficulties can jeopardize a country's development prospects. An effective energy policy for a developing country must consider two necessary responses, one short term and one long term. The short term involves adjusting to sudden increases in oil prices. Because oil imports tend to be a significant part of total imports, and because it is difficult to reduce oil consumption or non-oil imports in the short run when a major discontinuity in oil prices occurs, the increase in oil prices leads to a sizeable escalation in total imports. In the absence of an equivalent rise in exports, trade balances deteriorate and the resulting deficits have to be financed by borrowing or running down foreign exchange reserves.

SUBSTITUTION PROBLEMS

In simple terms, the key issue for OIDCs in the long run is to find alternatives to oil imports. This, however, is not easily accomplished. To effect such a change in the pattern of energy utilization, developing countries will have to make fundamental structural adjustments to their economies. This transformation is particularly difficult given the links among energy prices, income distribution effects and environmental policy goals. Policies designed to accomplish one objective may be offset by the intractable nature of other societal goals. For example, while at one level it is clearly a major policy goal for most LDCs to ensure that all their citizens have access to affordable energy, such a policy prescription may work at cross purposes with measures designed to foster the rapid development of alternative fuels or to bring greater economic rationalization to the economy.

Clearly, the immediate effect of levying higher energy prices is to reduce the incomes or profits of energy users. However, policy problems arise in that to the extent higher prices cannot be passed on, energy users will have little choice but to substitute less expensive fuels for more expensive ones, to improve energy inefficiencies, or to reduce consumption levels.

While it is difficult to implement these structural changes on a timely basis even in an advanced industrial economy, in most OIDCs it is much harder to substitute labor, capital and other raw materials for energy, owing to a host of infrastructural problems (a shortage of skilled technicians and managers or the infrastructure needed to utilize alternative fuels).

Each OIDC will have to implement an energy program commensurate with an analysis of its own problems, prospects and priorities. Only when such an assessment has been made can the government of an OIDC chart out a long term adjustment program.

The difficulties confronting OIDCs are extremely complex. Although many OIDCs do not yet have capital stocks and infrastructure that are as energy intensive as those in the OECD nations and consequently may be able to build new industrial facilities that utilize energy more efficiently, they also possess limited capital resources and experience balance of payments constraints that foreclose policy options that would otherwise be available to them. These constraints make it extremely costly (both economically and politically) for Third World leaders to divert resources from other needy economic sectors (health, medicine, housing, agriculture, etc.) into developing a more energy efficient capital stock.

With the emergence of the 'oil glut', and softening real oil prices, the choices confronting OIDC policy makers have become even more fraught with uncertainty. While many analysts believe that the long term price of oil will increase at a real annual rate of 2-3 percent, this is by no means assured. Indeed, if the market behaves as it has in the past, future price increases may come abruptly, followed by gradually declining real prices. In the past, many countries, in responding to such price behavior, failed to make the necessary adjustments in 1973-74, i.e. reducing their oil imports. This left them unprepared for the abrupt price rises of 1979-80. Currently the market is such that there is considerable downward pressure on oil prices. This could continue, or the market could tighten owing to a variety of factors (one being increased demand in response to lower prices). Although assumptions concerning future energy prices are at best tenuous, an OIDC which avoids the immediate benefits of importing lower priced oil and pursues policy measures designed to reduce oil imports in the long run may be embarking on a more prudent course.

However, the benefits of this policy course in the long run may exacerbate a country's short to medium term problems. While a critical ingredient of effective economic planning, dynamic energy policies are successful only to the extent that they promote broader economic, political and social development goals. In this regard, while it is essential that energy concerns play a major role in the enactment of all policies towards the industrial, agricultural, rural and transport sectors, countries should not automatically forgo the development of certain economic sectors simply because they are energy intensive. Two cases in point are (1) modern methods of farming which while more energy intensive per hectare generally use less energy per unit of output and (2) energy intensive industries such as aluminium, copper refining, fertilizers, iron and steel. In the latter case, though these industries consume large quantities of energy, they also often provide substantial tax revenues and/or large export earnings. While it is difficult to generalize for over 100 countries, countries possessing cheap hydroelectric power or non-exportable natural gas reserves may find that the development of energy intensive industries gives them a comparative cost advantage in international or regional markets.

EXTERNAL ASSISTANCE

External development assistance will continue to be necessary owing to the inability of most OIDCs' national economic institutions to finance new energy supply and development projects. This is especially true of the private sector of many of these nations where interest rates from private lenders are often much higher than those from public agencies. Low government rates are usually justified on the grounds that the projects being financed are meant to benefit the public as a whole. However, in such a case, although the apparent financial cost of a project is reduced, the real cost to the public at large may be substantial. Clearly, private lending institutions cannot afford to behave in such a manner. Another constraint is the fact that many countries have been slow to develop well-organized capital markets. Consequently, financial institutions have not developed to the point of bringing savers together with borrowers who wish to undertake a certain project. Where this occurs as a result of a lack of adequate competition, innovative banking procedures (conducting local risk assessment, and establishing decontroled decision making) can help to broaden the capital markets. Despite these efforts, however, the financing needs of large energy development projects will not usually be met through local institutions.

Let us now turn our attention to possibilities of financial assistance from foreign sources. The types of external sources of energy development assistance which will play relatively important roles in the 1980s are as follows:

1. financial assistance from official institutions;
2. financial assistance from private banks; and
3. foreign investment.

In regard to lending from official creditors, the trend has been one where the share of official debt has been declining relative to the total debt of developing countries. This trend, likely to continue in the 1980s, will have a negative impact on the low-income developing countries, which have not shared in the benefits of increasingly large private international capital flows. A substantial portion of the borrowing has been done by the higher income countries and a few middle income countries. In the second quarter of 1979, for example, six middle income countries accounting for 63 percent of all Eurocurrency credits extended to developing countries. The increased share of private lending has been the principal reason for the hardening of terms of the OIDCs' outstanding debt. At the same time, the emphasis placed on the role of private capital by the Reagan administration will accelerate this trend.

In the past, official and private capital have tended to complement each other, rather than substitute one another:

> 'While private capital is largely untied, of comparatively short maturity, and (increasingly) at floating interest rates, official capital is more often tied (by country of expenditure and by project), of longer maturity, and mainly at fixed interest rates. . . The involvement of official capital in infrastructure and other long-lived projects, where social returns are high, but difficult to appropriate, has had the effect of leaving more "bankable" projects . . . for private investors and of raising the returns of these projects.'

However, as private capital increasingly substitutes for official capital infrastructure and social service projects may have to be forgone, or debt service problems may be exacerbated (owing to the high capital intensity of such projects).

THE ROLE OF THE INTERNATIONAL AGENCIES

This situation could be eased by a significant increase in official lending. However, prospects for this appear dim. The World Bank's current programs for loans and grants to OIDCs for energy related projects amount to $13.2 billion through 1985. While the IBRD has proposed doubling its lending authority and has supported the creation of a new energy affiliate, the Reagan administration's opposition to it has effectively nullified the proposal.

Still another IBRD activity which could provide increased capital flows to help LDC energy development is the International Finance Corporation. Under guidelines established by the Bank, the IFC can serve as an intermediary bringing developing country governments together with foreign investors in order to finance energy projects. The IFC has been particularly successful in situations where companies have confirmed a discovery but are reluctant to commit further funds until contract terms are agreed upon. In such a case, IFC's involvement is to facilitate negotiations to bring the parties together. IFC will also assist OIDCs in locating a foreign investor if a government wishes to increase its refinery capacity and prefers a foreign company to undertake the task. The corporation will also be a participant in financing ventures involving secondary recovery programs. In Zaire, for example, the IFC is investing $4 million in a $33 million project in an offshore oilfield.

These types of arrangements, i.e. co-financing and official participation in foreign investment, could be substantial vehicles to assist

OIDCs in their adjustment problems during the 1980s. The US administration would do well to continue supporting policies geared to promote international trade and investment. In order to strengthen the viability of co-financing, strong support for the functions of the World Bank and its role in international development is advised. Contrary to the perceptions of the Reagan administration, this would not mean a reduction in the role of the private sector. Such activity could perhaps enhance the role of the private sector by providing outlets for its funds which otherwise might be foreclosed. Additionally, such arrangements could serve to increase capital flows to the OIDCs.

In addition to the work of the IBRD, OIDC access to funds provided under IMF facilities has been expanded allowing the Fund to make a significant contribution to the OIDCs' financing needs during the adjustment period. Member countries in severe payment difficulties may receive up to 150 percent of quota annually or 450 percent of quota over a three-year period, up to a cumulative limit of 600 percent of quota, subject to the Fund's conditions. In recent years, members have been more willing to seek the Fund's assistance in financing and implementing adjustment programs. In the past two years, about three-fourths of the finance from regular Fund resources has been made available under terms of high conditionality. This reflects the large scale balance-of-payments deficits of recent years and the recognition that such imbalances require stronger adjustment programs. In the financial year ending 30 April 1981, OIDCs borrowed, under various arrangements, about 6.6 billion SDRs. The Fund is currently undertaking arrangements to provide supplementary resources of 6–7 billion SDRs over the next two years in anticipation of continued balance-of-payments difficulties. While this is a major substantive step in alleviating OIDC short term deficit problems, many OIDCs use the IMF only as a last resort owing to its tough terms of conditionality discussed earlier in the paper.

OPEC AID FLOWS

As a result of the large balance-of-payments surpluses the OPEC nations enjoyed in the wake of the 1973–4 oil price increases, the international community placed great pressure on OPEC to expand the volume of its international aid. Partially in response to such pressure, and partially owing to its self-interest in maintaining an identity with other developing nations, OPEC increased its aid flows from $1.3 billion in 1973 to between $7–8 billion in 1980. OPEC countries (primarily the Arab countries) have been giving an average of $4–5 billion dollars per year in official development assistance to OIDCs through a variety of bilateral and multilateral arrangements. This aid, as a percentage of GNP, has compared favorably to the aid programs of the OECD countries; in 1980, it was estimated at 1.36 percent, compared to 0.37 percent for the OECD countries. An important part of this aid is earmarked for the development of energy sources in OIDCs, as well as lending for other non-energy related development projects.

However, it is not likely in the wake of falling oil prices that OPEC aid will increase in size or substance during the 1980s. With many OPEC countries having large populations and relatively low oil reserves (Algeria, Iran, Nigeria, Venezuela, Iraq) now in current account deficit, they are more likely to focus on their own development needs. While Saudi Arabia, Kuwait and Qatar are in a position to maintain or increase the level of their contributions if geopolitical factors within the Middle East continue to dominate the concerns of the wealthier nations, the primary recipients of such aid will be confined to neighboring Middle Eastern countries.

Moreover, OPEC has resisted the suggestions of the OECD nations to (1) sell oil to OIDCs at a lower price, and (2) to increase direct investment to OIDCs. From the OPEC perspective, the first suggestion opens the possibility of low-priced oil leaking into other markets. It would be difficult to establish safeguards to avoid such an occurrence, especially in the case of developing countries that are refiners and that have reexport capacity in refined products. The OPEC countries prefer to extend soft loans to OIDCs to cover periods of balance-of-payments difficulties. In regard to direct investment, OPEC considers such a move as unwise. OIDCs have poorly developed infrastructures, immature money markets, and uncertain attitudes and policies towards foreign investment. OPEC views its surplus funds as a means for investing in its own development plans. Thus, although the official aid coming from OPEC has been significant, future increases in OPEC's role as a lending institution will be moderate.

OECD INVESTMENT AND LENDING

The OECD nations have not matched the OPEC nations in their lending as a percentage of GNP; volumetrically they have surpassed it. While many OECD nations have made a commitment to increase their ODA, the new US administration has exhorted developing countries to rely less on foreign aid, and more on international trade and foreign investment. The direction which ODA will take is therefore unclear. Although increases by other OECD nations may offset decreases in aid by the US, this appears unlikely. In the authors' opinion, the new initiatives of the US administration will achieve

The OECD nations have not matched the OPEC nations in their lending as a percentage of GNP; volumetrically they have surpassed it. While many OECD nations have made a commitment to increase their ODA, the new US administration has exhorted developing countries to rely less on foreign aid, and more on international trade and foreign investment. The direction which ODA will take is therefore unclear. Although increases by other OECD nations may offset decreases in aid by the US, this appears unlikely. In the authors' opinion, the new initiatives of the US administration will achieve mixed results, the net effect of which is still unclear. The International Development Association has wrongly come under attack by the administration as being an agency for handouts. The World Bank has historically used prudent procedures and criteria for project appraisals, resulting in a limited number of sound projects receiving financial support. These projects have generally resulted in the Bank receiving a profit. In light of this, it is unfortunate that the administration has refused to support the proposal for the establishment of a separate energy affiliate, which would certainly be professionally managed. On the other hand, the stress on the use of private capital for development projects and the defense of international trade against protectionism could increase the role of private lending institutions and foreign investment, as well as enhance the export markets of the OIDCs if global economic recovery occurs soon.

The role of direct foreign investment in regard to international capital flows to OIDCs actually decreased during the 1970s as a result in large part to restrictive regulations of host country governments with respect to foreign investment. Another factor contributing to this decline has been the fear of many multinational corporations of expropriation. Domestically, governments have either actively provided disincentives to go abroad or have adopted a lackadaisical approach toward export promotion. Nowhere has this lack of commitment toward export promotion been more apparent than in the United States.*

*A further 12 pages of this paper are devoted to specific opportunities in the United States.

Chapter 23

The Scope for Adjustment

Lutz Hoffman and Lorenz Jarass†*

INTRODUCTION

There is now considerable confusion about the impact that rising oil prices had in the past on oil importing developing countries. Whereas on the one hand rising balance of payments deficits and high debt servicing ratios are quoted as indicators of a disastrous impact, it is argued by others that economic growth of these countries has hardly suffered during the second half of the 1970s and that the worsening of the balance of payments situation is largely due to other factors than the rise of oil prices.

Unfortunately, there has been little solid analysis of what the impact of rising oil prices on oil importing developing countries can be and whether and to what extent these countries can adopt adjustment measures which help to largely mitigate the impact. This paper reports about a research project which tried to answer these questions by use of a simulation model. The model was applied to three countries: Brazil, India and Kenya. By choosing three rather

*Professor, Department of Economics, Regensburg and Chairman of the German Chapter of the International Association of Energy Economists.

†Dr Jarass teaches at the University of Regensburg and is Managing Director of the research association, ATW.

different countries it was hoped to derive conclusions which could be generalized for a wider range of different oil importing developing countries.

In the first section we give a verbal description of the model. A formal description is provided in the appendix.* The second section* reports about the various simulation runs made with the model, while the third section summarizes the major conclusions.

MODELLING THE IMPACT OF RISING OIL PRICES

An exogenous increase in oil prices affects the economy via its impact on final as well as intermediate demand for oil products. If oil products become more expensive the costs of intermediate products will also increase and this again leads to price increases for final products. This mechanism suggests that the impact of rising oil prices is best modelled within the framework of an input-output model.

In the long run one also can expect supply reactions to rising oil prices where the respective resources do exist. However, with long gestation periods for energy investments, such reactions would hardly materialize within a decade. Furthermore, up to now little is known as to what extent energy investments in developing countries depend on prices as compared to other factors, such as financing constraints, political stability etc. Supply reactions are therefore omitted in the present analysis.

A schematic overview of the model utilized for the impact analysis in the project under discussion is given in the attached flow chart. The model iterates as many rounds of adjustment as required, each of which can be considered as consisting of three phases. In describing these three phases it is best to distinguish between price reactions on the one hand and quantity reactions on the other.

1. Price Reactions

If the price of oil or oil products increases, the costs of commodities for which oil is used as an input rise in phase 1 (Fig. 23.1) according to the cost share of oil. For instance, if the cost share is 8% and the oil price increase 20%, costs increase by 1.6%. Does this mean that commodity prices increase by the same percentage? The answer is

***Omitted here but available in the BIEE Archive.**

positive, at least for the medium run, because one cannot expect that prices of other inputs go down. This is certainly the case for material inputs, but also for labour. Only the remuneration of entrepreneurs, i.e. profits, could temporarily decrease. However, in the medium run entrepreneurs will make sure that the profit margin they consider as an adequate remuneration for entrepreneurial services and incurred risk is maintained. Hence, a parallel movement of costs and prices is a reasonable assumption.

The impact of rising oil prices on costs could be mitigated by technological changes which substitute other inputs for the now more expensive oil input. There are two substitution processes to be distinguished. One, where oil is substituted for by other energy inputs and the other where non-energy inputs substitute for oil. The latter is only of relevance in the long run, whereas the first could already affect costs in the medium run. However, the cost reducing effect of the first is relatively low compared to the latter, if prices of other energy sources increase also, though may be to a lesser extent than the oil price. Due to lack of sufficiently reliable information it was not possible to take substitution effects into account in the model presented here.

After the price increase of the first round producers see their input prices inflated and adjust output prices again in order to bring them in line with increased costs. This theoretically continues over an infinite number of rounds, though in most cases the additional price increase becomes negligible after five to six rounds.

The commodity price increases change the cost structure of the various producers. That means that the input-output matrix which is in value terms also changes in every round. The model therefore continuously adjust the input-output matrix in phase 3 according to the changing price structure.

The oil price increase leads to further general price increases in phase 3 if wage earners attempt to compensate for the loss in purchasing power by demanding higher wages. Whether such consecutive wage increases take place or not depends on institutional factors, for instance the existence of powerful labour unions, and political decisions. The model therefore defines wage increases as a policy variable and simulates the outcome of alternative wage policies.

The price increases for domestically produced commodities are assumed to be independent of whether the good is sold on the home market or exported. The impact of a strategy which differentiates between a (higher) domestic price and a (lower) export price can, however, be deduced from a simulation run presented below, where the value of exports is assumed to be constant. This case can be interpreted as either a situation where export prices remain constant

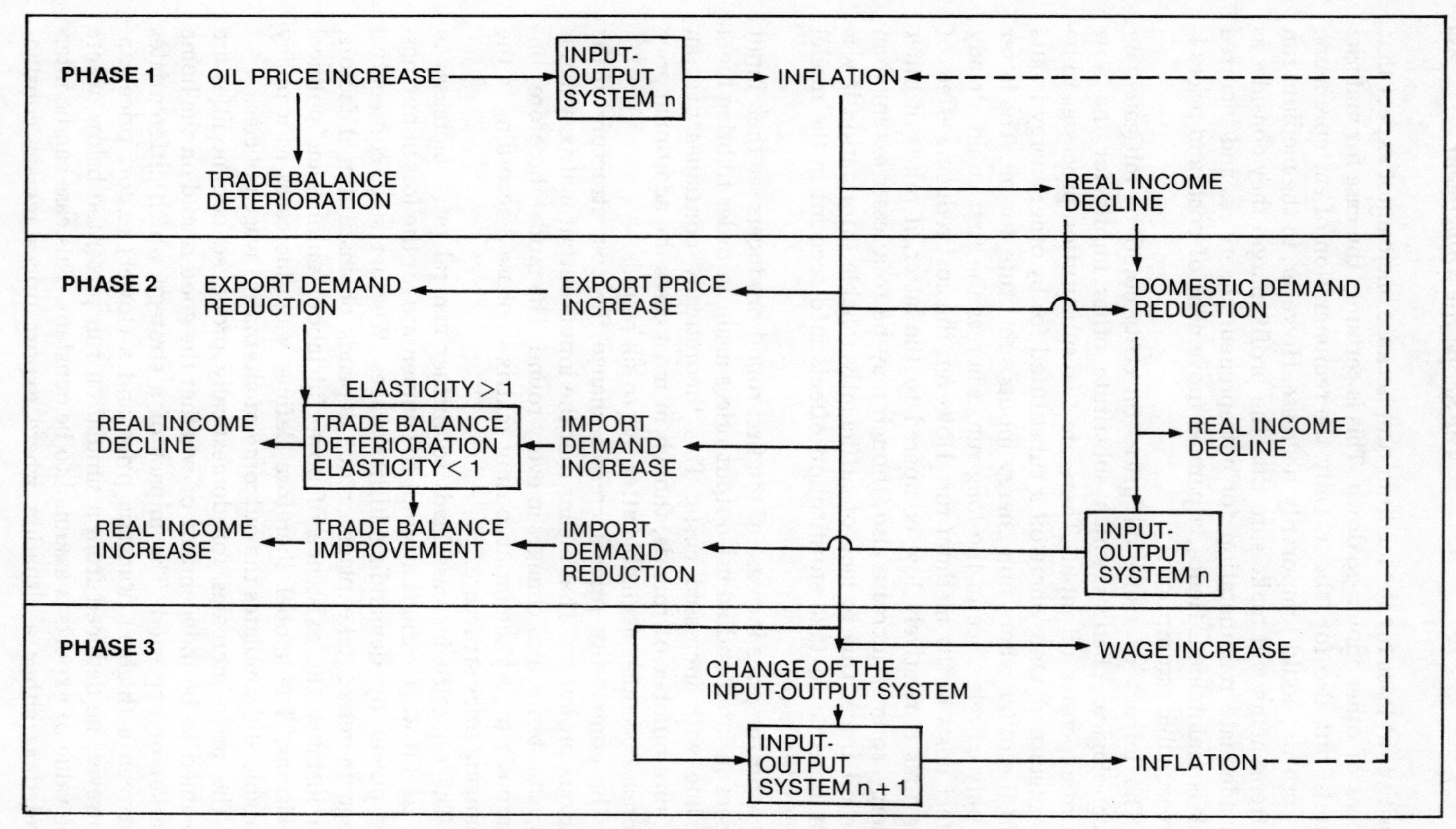

Figure 23.1. Flow chart of the simulation model.

inspite of rising domestic prices or one where the price elasticity of export demand is unity.

For import prices it was equally assumed that the price increase of a particular import commodity corresponds to the one of the respective domestic commodity. This makes sense if technologies do not differ significantly between countries and the costs of a commodity in one country are therefore affected by rising oil prices in a similar way as in another country.

2. Quantity Reactions

The price increase in phase 1 immediately reduces real income in two ways (see Fig. 23.1). First, the trade balance worsens and, second, the purchasing power of money income diminishes.

The decline in real income reduces in phase 2 the demand of the various commodities according to the respective income elasticities. Demand is further affected by changes in real prices. As the real price change is the increase of nominal commodity prices deflated by the general price index, some real prices rise whereas others fall. The model captures the combined impact of changes in real income and real prices by a log-linear consumption function (see Appendix).

The decline in real income affects imports in the second phase in two ways. First, imports for final demand decline in a similar way as domestic demand of the respective commodities. Second, the fall in domestic demand reduces the import of inputs via the input–output system. Both counteract the balance of trade deterioration of phase 1.

The price changes as such can affect the balance of trade positively or negatively. Assuming normal (negative) import demand elasticities, an increase in real prices reduces import demand and improves the balance of trade if the import demand elasticity is greater than unity. A worsening of the balance of trade can either come about if the elasticity is smaller than unity or if the real price decreases for commodities with an elasticity of larger than unity. The direct impact of price increases on the balance of trade is therefore ambiguous.

The oil price induced rise of export prices improves the balance of trade if export demand is inelastic. This assumption is plausible if the oil price increase affects all suppliers in all countries in the same way and the compètitive position of a supplier in a particular country is therefore not changed. The export demand will then only decline to that extent as the general price increase for that particular commodity reduces world demand. If the price elasticity of world demand is greater than unity, the value of export is reduced and the balance of trade worsens.

The remaining two components of aggregate demand, government consumption and investments, are largely determined by political decisions. Government expenditure which appears in the public budget usually is the result of a political compromise, and decision makers therefore may not want to change it if prices rise faster than originally assumed. Such a situation would correspond to the assumption of a price elasticity of minus one for public expenditure. It is, however, conceivable that public authorities are determined to execute the budget in real terms. That means, nominal expenditure has to be adjusted to rising prices. This would imply a price elasticity of zero. Both assumptions, a price elasticity of minus one and one of zero, are used in the simulations in order to determine the range in which the impact of different expenditure policies lies.

3. Dynamic Properties of the Model

Following the traditional approach of input-output analysis one would try to determine the new equilibrium price vector after an oil price increase. If $<\pi>$ is the vector of primary input prices including oil, A' the transpose of the input-output matrix and $<p>$ the new equilibrium price vector, the calculation runs as follows:

$$<p> = [I-A']^{-1} <\pi>$$

The implicit assumption of such a calculation is, that the new equilibrium price vector materializes instantaneously. This obviously is unrealistic. We assume, that the adjustment of commodity prices to a rise in oil prices stretches out over time and approaches equilibrium only gradually. As theoretically equilibrium is reached after an infinitesimal number of rounds it is in practice never reached and all observable price vectors represent disequilibria. We have allocated two rounds to each year, so that after a ten year period twenty rounds are covered. By then, one would, however, be very close to equilibrium if there were only one oil price increase at the beginning of the period.

This is, however, not the case. It is assumed that the oil price rises either every year at a certain rate or at two points of time within the ten year period in a shock-like upwards movement. In both cases the disequilibria originating from one price increase come on top of those from an earlier price increase. This is what actually happens in reality and why the input-output matrix is continuously adjusted according to the changing price vector.

Another dynamic property is the determination of profits. Entrepreneurs are assumed to calculate profits on the basis of input prices of primary factors in the present period and input prices of intermediate products in the previous period. The reason for this distinction is that entrepreneurs in the present period do not yet know what prices of intermediate products will finally be. They therefore expect intermediate product prices to be the same in this period as in the previous one. One could also assume that the expected prices of intermediate products are a function of past price changes. In the simulations it turned out that this assumption does not make much difference as long as one does use an extreme specification.

The residual character of profits results in strong fluctuations of this income category. After an oil price increase, profits are first strongly squeezed but then recover during the adjustment process. Whether the income distribution is changed in the medium run depends on wage policy. This again is what can be observed in reality.

The consumption function used in the model, finally, is dynamic in the sense that the change in consumption is related to the previous year's change in real income and real prices (see Appendix). There is one consumption function for every commodity group. The various functions are log-linear and fulfil the adding-up conditions. The constancy of elasticities implied by this specification does not appear to be unduly restrictive for a simulation period of ten years.

SUMMARY AND CONCLUSION

Table 23.1 provides an overview of the results obtained for a 3% oil price increase under alternative assumptions. It is seen that real income declines most and inflation is highest if nominal wages are adjusted in line with the rate of inflation. The lowest impact on real income is obtained if government expenditure and investment are not eroded by inflation but maintained in real terms. The other side of the coin is that in the first case (full wage adjustment) the balance of payment deterioration is least while it is highest in the latter.

A comforting result is that alternative assumptions about income elasticities and price elasticities do not significantly affect the outcome of the various simulations. Hence, we need not worry very much whether the parameters utilized by us correctly reflect the situation in the three countries. Only with the *mutatis mutandis* approach do different income elasticity values matter. However, even there our results should reflect the right orders of magnitude as the deviation of the actual parameters from the ones used certainly is smaller than the difference in parameter values of the two respective simulations.

Table 23.2, finally, shows cumulative effects of alternative constellations. It is seen, for instance, that the impact of wage policy very much depends on whether and to what extent public expenditure

Table 23.1. Summary of Results for 3 Percent Average Increase of Real Oil Price

Country	Run	Annual Increase	Shock	Public Expenditure & Investment real const.	Non-oil Export nominal const.	Wages real const.	All Income Elasticities	Price Elasticities	Mutatis Mutandis
	Y_r	−0.43	−0.45	−0.27	−0.47	−0.84	−0.44	−0.42	−0.49
Brazil	p	0.32	0.34	0.32	0.32	1.07	0.32	0.32	0.32
	Im–Ex	1.18	1.40	1.55	1.54	0.17	1.21	1.07	–
	Y_r	−0.19	−0.20	−0.14	−0.20	−0.35	−0.20	−0.18	−0.28
India	p	0.07	0.08	0.07	0.07	0.36	0.07	0.07	0.07
	Im–Ex	0.91	1.03	1.04	0.99	0.68	0.94	0.87	0.92
	Y_r	−0.48	−0.51	−0.34	−0.67	−0.71	−0.51	−0.47	−0.64
Kenya	p	0.42	0.45	0.42	0.42	1.22	0.42	0.42	0.42
	Im–Ex	0.45	0.52	1.27	1.24	−0.96	0.54	0.41	0.37

Table 23.2. Cumulative Impact of a 3 Percent Real Oil Price Increase

R: Constant in real terms N: Constant in nominal terms

Public Expenditure and Investment		*N*				*R*			
Wages		*N*		*R*		*N*		*R*	
Non-oil Export		*N*	*R*	*N*	*R*	*N*	*R*	*N*	*R*
		(1)	*(2)* *Base case*	*(3)*	*(4)*	*(5)*	*(6)*	*(7)*	*(8)*
Brazil	Y_r	−0.47	−0.43	−1.00	−0.84	−0.32	−0.27	−0.36	−0.19
	p	0.32	0.32	1.07	1.07	0.32	0.32	1.07	1.07
	Im–Ex	1.54	1.18	1.40	0.17	1.90	1.55	2.54	1.43
India	Y_r	−0.20	−0.19	−0.40	−0.35	−0.15	−0.14	−0.17	−0.13
	p	0.07	0.07	0.36	0.36	0.07	0.07	0.36	0.36
	Im–Ex	0.99	0.91	0.97	0.68	1.13	1.04	1.30	1.30
Kenya	Y_r	−0.67	−0.48	−1.24	−0.71	−0.52	−0.34	−0.68	−0.18
	p	0.42	0.42	1.22	1.22	0.42	0.42	1.22	1.22
	Im–Ex	1.24	0.45	1.08	−0.96	2.01	1.27	2.91	1.21

and investment are adjusted. Without adjustment a wage policy that maintains real wages has a rather negative effect on real income, whereas with adjustment the effect of such a policy on real income is positive. The balance of payments effect always goes in the opposite direction.

One may conclude that there is no dominant solution in the sense that for all three targets evaluated one parameter constellation is superior to the others. Which constellation is to be chosen in a certain situation depends on the relative weights the government attaches to the three targets.

PART XVI

Energy Prices and Global Growth

Abstracts (see Section 5)

Chapter 24

Population, Food, Energy and Growth

*Wassily Leontief and Ira Sohn**

The purpose of this paper is to present a general background for and a preliminary inquiry into the longrun prospects of providing for the growing population of the world with emphasis on two principal components of economic well being, food and energy.

To think in global terms does not mean to view the world as a single, undifferentiated entity. On the contrary, one has to visualize it as a complex system of many different but nevertheless interrelated and interdependent areas. The growth or the stagnation of each one of them will to a great extent depend in the future, as it did in the past, on what happens in the others. The same of course is true of each individual national economy: each one of them consists of many different sectors, but none can function (except under very primitive conditions) without utilizing as inputs commodities and services produced by other sectors and delivering its own output to other sectors. To explain these operations and to understand the development of a national economy, one has to describe and analyze it as a system of distinct but mutually interdependent activities. A systematic procedure starting with a detailed description of the input–output structures of the individual sectors of each region and

*Institute for Economic Analysis New York University.
†The full paper with methodology and results is available in the BIEE Archive.

then proceeding on that basis to the analysis of the economic interdependence among various regions within the framework of the global economy as a whole offers a convenient tool for carrying out such a task. In raising questions concerning the unknown future it seems to be appropriate to base our conjectures on a concrete assessment of structural changes that can be expected to affect the production and consumption of specific goods and services in different parts of the world.

The set of alternative projections of prospective growth of the developed and less developed countries presented in this paper is derived on the basis of the large multisectoral, multiregional input-output model of the world economy constructed for the United Nations (*The Future of the World Economy*, Oxford University Press, New York, 1977) several years ago.

The original projections covered the thirty-year period 1970-2000. In advancing the time horizon from the year 2000 to the year 2030, the analytical structure of the model has been retained with its emphasis on the mutual interdependence of the sixteen regions and between some forty different branches of production and consumption within each of them. The numerical database of the original model describing the present input-output structure, that is, the technological characteristics of various agricultural and industrial sectors and their prospective changes, has to a large extent been retained too.

The limited resources available for revision of the base figures were devoted mainly to refining and improving the description of the technical characteristics in light of the latest available information and expert judgment. Prospective changes in the input-output structure of the various branches of agriculture and the structural relationships controlling the production and consumption of principal kinds of energy and non-fuel minerals were also reassessed for this study.

Each one of the four alternative projections of the future growth of the nine developed and the seven less developed regions into which the world economy was subdivided is internally consistent. That is, the total world output (production) of each good and service equals its aggregate world input (consumption), and the combined regional exports of each internationally traded good are equal to its combined regional imports. Moreover, the composition and allocation of the domestic output, exports and imports of each good, are made to satisfy domestic consumption and investment requirements in accordance with the prevailing state of sectoral technologies. The level and composition of (physical) investment in each region are determined so as to provide as time goes on new productive capacities corresponding to the requirements of the projected economic growth.

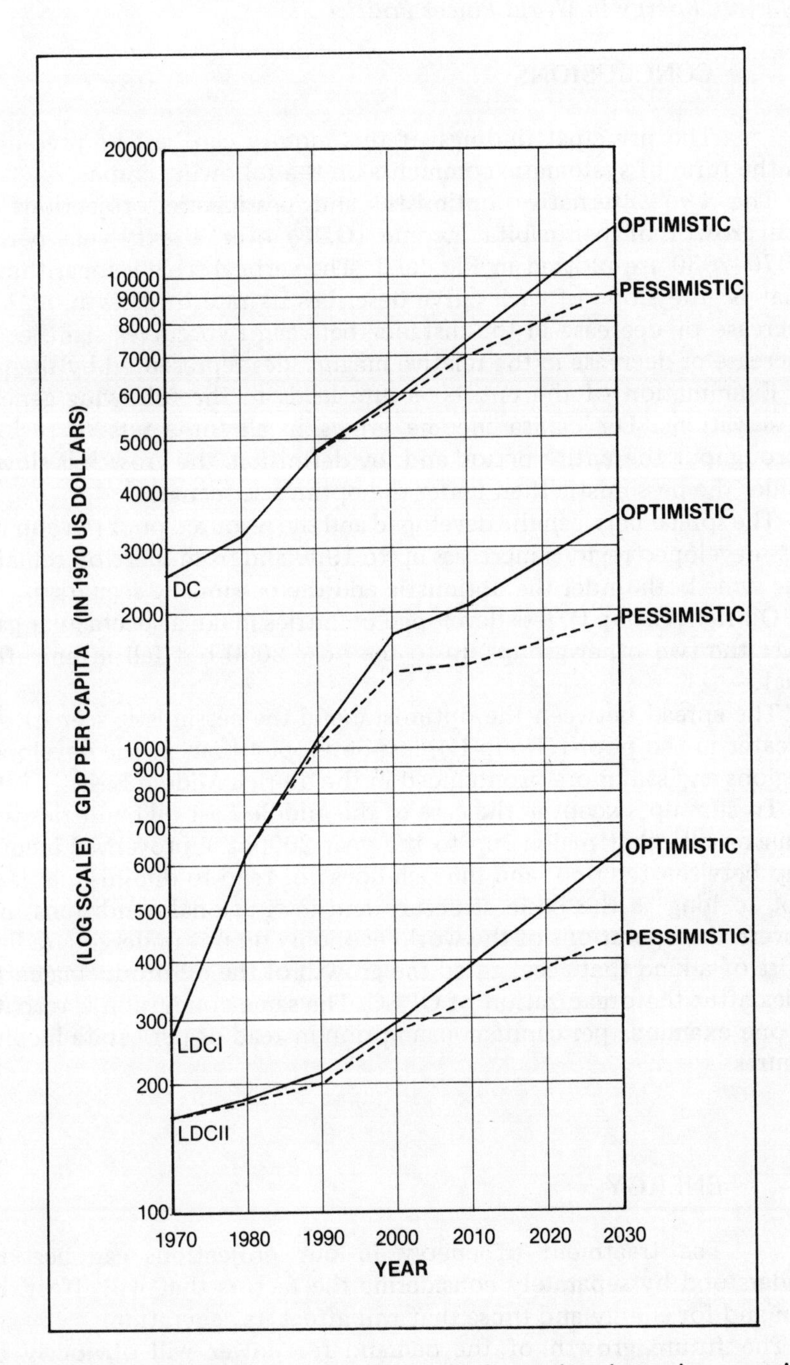

Figure 24.1. Gross domestic product per capita under alternative scenarios. Key: DC = developed countries; LDCI = developing group I (resource rich); LDCII = developing group II (resource poor).

CONCLUSIONS

The principal findings of this inquiry can best be presented in the form of systematic comments on the following graphs.

The two alternative optimistic and pessimistic projections of the growth of per capita income (GDP) over a sixty-year period 1970–2030 are plotted in Fig. 24.1. The vertical scale is logarithmic, that is, the slope of each curve describes its rate of growth, and an increase or decrease in the distance between two curves signifies an increase or decrease in the relative magnitudes represented by them.

Examination of the entire picture leads to the following general observations. Per capita income grows in all three sets of regions throughout the entire period and, by definition, the growth is slower under the pessimistic than under the optimistic scenario.

The spread between the developed and the resource-poor (Group II), less-developed regions increases up to 1990 and from there on remains the same both under the optimistic and the pessimistic scenarios.

Oil-rich (Group I), less-developed countries make a spectacular gain over the two other groups up to the year 2000 but fall in line after that.

The spread between the optimistic and the pessimistic scenarios is greater in the poor (Group II), less-developed than in the developed regions and still more pronounced in the oil-rich Middle East.

To sum up, except in the case of the Middle East oil countries that gain on all other regions up to the year 2000, the proverbial income gap between the poor and the rich does not tend to diminish, at least not so long as the basic structure and institutional conditions that govern the operations of the world economy do not undergo a radical shift of a kind that accelerated the growth of the oil-producing countries after the organization of OPEC. The same conclusion is reached if one examines per capita consumption instead of per capita income figures.

ENERGY

The treatment of energy in our projections can best be understood by separately considering the factors that will affect the demand for energy and those that will affect its generation.

The future growth of the demand for power will obviously be greatly affected by the adoption of energy saving technologies. Within the framework of the input-output system this process is described concisely by a gradual reduction in the technical coefficients that make up the structural matrices which constitute the common database for all of these computations.

On the supply side special attention was given to incorporating in the same database new sets of input coefficients that reflect the technical structure of the energy producing sectors. Oil, natural gas, hydroelectric and nuclear power are expected to provide in the next fifty years an overwhelmingly large part of the total energy supply both in the developed and in the less developed countries. Even under the optimistic scenario the bulk of it will be produced and consumed in the six industrialized regions.

Nuclear and hydroelectric energy is being employed only in the form of electric power, while fossil fuels are transformed in part into electric energy and at the same time also used directly. While oil and natural gas can be expected to play a diminishing role because of gradual exhaustion of limited reserves, additional energy demand will have to be met by increased output of coal—mainly American coal—or of atomic power. Computations show that so far as the overall picture of economic growth is concerned, the choice between the two makes little difference.

The choice between atomic energy and coal is bound, however, to have a marked effect on the internal structure of the industrialized developed economies as well as the trade relationship between them. In this context the distinction between the coal and the nuclear versions of our pessimistic scenarios becomes quite significant. Since constructing nuclear plants as well as opening new mines and providing transportation facilities to move their output take much time, the split between nuclear and coal versions of the optimistic and the pessimistic scenarios was introduced in our projections only after the year 2000.

The overall intensity of energy utilization in any particular economy can be measured as the amount of energy used per unit of GDP. The curves plotted in Fig. 24.2 show how these energy use ratios would move, according to our projections, in each of the three groups of advanced industrial countries under the alternative coal and nuclear scenarios. The picture is practically the same for the optimistic and the pessimistic scenarios.

In all regions and at all times less energy would be produced and used per unit (1970 US dollars) of GDP under a nuclear than under a coal regime. This probably is due to the great amount of power absorbed in the mining of coal and moving it to the power plants. As time goes on the efficiency of energy use rises under the nuclear regime in all regions both under the optimistic and the pessimistic scenarios. This is not so, however, under the coal regime. This can be explained by the fact that the introduction of nuclear power—which

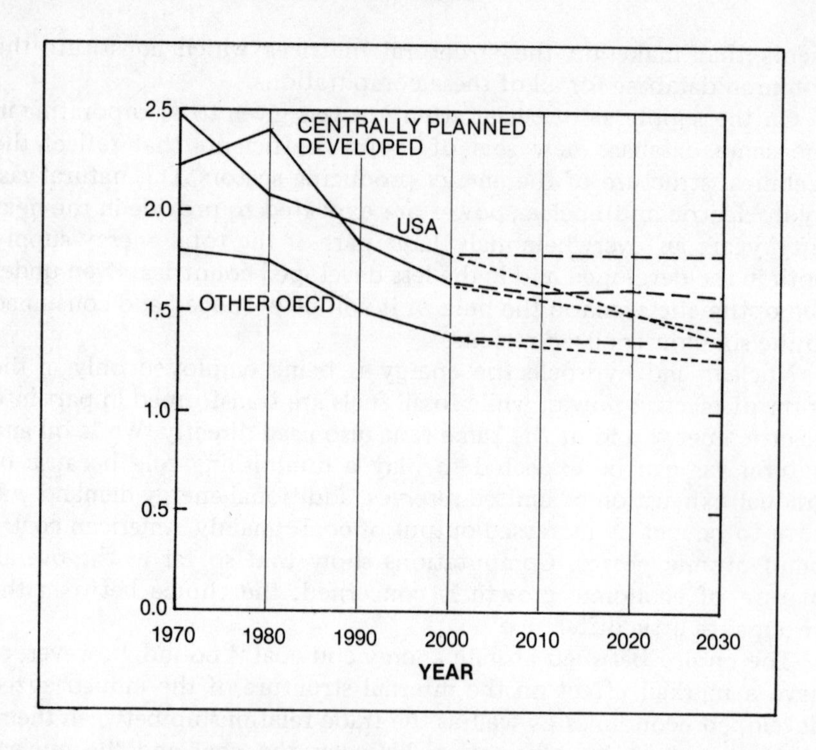

Figure 24.2. Total energy use per dollar of GDP under alternative scenarios. Key: — base trend; – – – coal scenario; · · · · nuclear scenario. Note: energy use measured in million metric tons of coal equivalent; GDP measured in 1970 US dollars.

seems to be on balance more efficient than coal power—is a drawn out process which affects the energy balance of developed economies gradually step by step.

The amount of capital that has to be invested per unit of output of electric energy production in an atomic plant is much larger than in a plant using fossil fuel. However, their difference in total capital requirements may not be as great if one considers the investment needed in coal mines and railroads or other means for delivery of the fossil fuel to the generating plants. As a result, even if GDP for each developed region is determined endogenously, we would expect similar projected growth rates in the coal and the nuclear scenarios.

Chapter 25

The International Energy Investment Dilemma

*Paul Tempest**

There is today no such phenomenon as a unified international energy market. It is a myth—a very dangerous myth. Let us dispose of it at the outset.

There never has been a global energy market. What there is, at the consumer end, is an amalgam of national energy systems developed from local resources on to which has been grafted various energy imports. Of these the only truly internationally traded energy commodity is imported oil. For their base load energy needs, most national economies have relied on committed systems of domestic supply usually embedded deeply in their own public utility infrastructure and industrial system. The prices to the consumer of all these forms of energy are heavily masked by subsidy, taxation and other forms of government control. The availability of domestic energy varies greatly from country to country. There are therefore great variations between countries (and also within countries) in the resource cost of energy and in the price to the consumer.

*In compiling these remarks, the author has drawn on the reports of the 21 Session Chairmen and 19 Rapporteurs of the 1982 Cambridge Energy Conference. He is particularly grateful to Dr D. F. Lomax, Group Economic Adviser of the National Westminster Bank, for assistance with the tables and diagnosis, to Finn Lied, Chairman of Statoil, for a vital part of the argument, and to Professor M. A. Adelmann of the Massachusetts Institute of Technology for his comments on the text.

No genuinely international market exists in the commodity natural gas. It is a collection of widely different bilateral arrangements in which the agreed price is but one component in the bilateral package. There is therefore no common world gas price and differences in import gas prices (whether fob or cif) are considerable. Imported coal, similarly, is a limited market: being dependent on heavy investment at the consumer end, it relies to a large extent on bilateral negotiation and committed long-term supply. The uranium market is also hardly a free market given the political and security aspects of securing long-term supplies.

As indicated above, there will be great variations between consumer countries in their ability and willingness to pay for new energy supply. The argument advanced strongly in recent years by several OPEC countries that gas (and coal) should be sold ex-producer at oil price parity (in thermal terms) is, as Professor Adelman describes it, 'a chimera' which has effectively halted much long-term energy contracting and other serious price negotiations. It is gradually becoming more obvious in a slack market that competitivity of a new energy resource is not at the producer end but 'at the burner tip', the consumer end, where it will only displace other energy forms if there is an advantage to the consumer to make the switch. Once this principle is fully accepted—and it is clearly difficult for politicians to jettison the rhetoric of a decade—the way is open to re-think a whole range of energy deals where the producer netback calculation, i.e. the buying price less the costs of transportation, still allows handsome net currency revenue for the producer.

This is only a first hurdle in breaking through the current impasse in developing new international energy projects.

THE NATURE OF THE DILEMMA

Oil, it has been generally assumed since 1973, is, and will be, a declining factor in global energy supply. Gas, coal and nuclear, which are expected to fill the supply gap, all require large integrated systems which bind the producer to the consumer for long periods of time. The 10, 15 and 20 year supply contracts common in these industries often assume investment lead-time of up to 10 years. The massive front-end investment necessary to produce a non-tradable product over a 20–30 year timeframe is quite different from standard oil industry financing. Its scale and character often demands the kind of backing and guarantee which only governments can give and the finance has to be raised, to a large extent on the international capital market, and through the world banking system.

The central dilemma of current energy investment financing on this scale is how to mobilise funds, in conditions of falling real oil (and energy) prices, global recession, ample current energy supply and surplus refinery capacity without incurring penal interest rates and other financial costs which express the anxiety and uncertainty of the investor.

THE PROBABILITY OF UNCERTAINTY

The only near-certainty in international energy over the next decade is that a wide area of *uncertainty* in energy supply, demand, price and potential is likely to persist.

There is at present a fundamental conflict between the short-term prospect of falling real oil price and a long-term consensus (based possibly on the perceived depletion of oil and gas reserves and more probably on the assumption that the oil producing and trading cartels will reform and reassert themselves) that the real oil price will have resumed an upward trend well before the end of the century. 'Talking the price down' can be seen as giving short-term relief to the global economy, whereas 'talking the price up' is preoccupied with the potential profitability of new investment. While both arguments hold their ground, decision-making in energy megaprojects, which in any case will not have an impact in supply terms before the nineties, tends to be postponed. Such delay further increases the chances of a third major energy price discontinuity, probably within the decade and probably sooner rather than later.

We need, therefore, to understand the nature of this prolonged uncertainty.

TOWARDS A DIAGNOSIS

The economic consequences of the major oil price hikes of 1973 and 1979/80 have now been well documented and widely analysed. The scale of the impact on the world economy has been so staggering that it has placed in question the whole financial mechanism of capital formation and long-term investment. One of the great ironies of the present situation is that, as the pendulum of energy price and economic activity swings back towards equilibrium, governments, international agencies, banks and industry in general seem more afraid of a price decrease than a further price increase. In the event of a third major oil price hike it is difficult to see how similar economic repercussions could be avoided.

Table 25.1
Growth of real GDP at market prices in the OECD area (Percentage changes)

	1972	1974	1976	1978	1980
United States	5.5	-0.6	5.4	4.4	-0.1
Japan	8.8	-1.0	5.3	5.0	4.4
Germany	3.7	0.5	5.2	3.2	1.9
France	5.9	3.2	5.2	3.7	1.2
United Kingdom	2.2	-1.0	3.6	3.3	-1.4
Italy	3.2	4.1	5.9	2.7	4.0
Canada	5.8	3.8	5.8	4.0	-0.1
TOTAL of above countries	5.3	0.3	5.2	4.0	1.2
TOTAL smaller European countries	5.3	3.6	3.8	2.5	1.9
TOTAL smaller countries	5.0	3.5	3.7	2.4	1.9
TOTAL OECD	5.3	0.9	4.9	3.8	1.3

Source: OECD

Table 25.2 Consumer prices (Percentage changes from previous year)

	1972	1974	1976	1978	1980
United States	3.3	11.0	5.8	7.7	13.5
Japan	4.5	24.5	9.3	3.8	8.0
Germany	5.5	7.0	4.5	2.7	5.5
France	6.2	13.7	9.6	9.1	13.6
United Kingdom	7.1	16.0	16.5	8.3	18.0
Italy	5.7	19.1	16.8	12.1	21.2
Canada	4.8	10.8	7.5	9.0	10.1
TOTAL of above countries[a]	4.3	13.3	8.0	7.0	12.2
TOTAL OECD[a]	4.7	13.4	8.6	8.0	12.9

Source: OECD
[a] The country weights used in the aggregate indices are based on the private consumption and exchange rates of the preceding year.

For developing countries, the ratio of oil prices to all other commodity prices increased over sixfold in ten years: a grotesque distortion of their ability to balance their external trade or to finance their own development. Heavy external borrowing has proved only a temporary palliative, and in turn imposed its own punitive burden of debt-servicing. In the OECD countries annual average real economic growth of 2.5% in 1973–80 was half the average annual rate in the previous 12 years, while the average rate of inflation more than doubled from 4.4% to 10.4%. Such averages mask vastly greater swings in individual countries (see Tables 25.1 and 25.2). One does not have to look far for a reason: mainly as a result of the increased bill for imported oil, which for the OECD rose from $20 billion in 1972 to $267 billion in 1981, the OECD current account swung violently into deficit in 1974, back into surplus in 1975, deficit in 1976–7, surplus in 1978 and deficit in 1979–80 (see Section 6, Table T). Deflationary measures to correct the deficits were one of the main causes of reduced growth; and OECD unemployment is now approaching 30 million.

The OPEC countries were also grappling with massive swings in their current account from a surplus of $1.3 billion in 1972 to $60 billion in 1974, back to $4 billion in 1978, then up to $105 billion in 1980 (see Section 6, Table S). They are now facing the prospect of a move into sharp net deficit. Investment planning in many of these producer countries has followed a cycle of grandiose development planning generating hyper-inflation followed by savage and wasteful retrenchment.

The volatility of exchange rates (see Table 25.3) and inflation, reflecting fluctuations in economic growth and the balance of payments, created great uncertainty in the financial markets, already struggling to absorb and recycle the huge surpluses of OPEC. In such circumstances, there was a steady drift towards higher real interest rates (only recently reversed) which acted as a further deterrent for investment and growth.

Expectations of further exposure to such violent swings in activity have prompted OECD policies of securing greater energy self-sufficiency, expressed at the extreme as ambitious megaprojects to stimulate domestic energy production, many of which have had to be later rationalized or axed. A parallel development has been the regionalization of falling energy demand which has cut OPEC production and exports sharply. Nonetheless, OPEC oil remains the key component of the energy picture and the fear of another random political event such as the October 1973 Arab/Israeli war, or the Iranian revolution—Iran/Iraq war, could have similar sharp effects on oil supply and price. Economic recession and a slack oil market

Table 25.3 Exchange Rates – Trade Weighted Index (1979 trade weights)

Date	*UK £*	*US $*	*DM*	*YEN*
18.12.71	100.0	100.0	100.0	100.0
6.7.73	81.63 (LOW)	89.7 (LOW)	121.0 (HIGH)	108.7 (HIGH)
31.12.74	78.24	93.8	118.5	95.3 (LOW)
31.12.75	71.74	98.5	116.0 (LOW)	97.7
29.10.76	56.83 (LOW)	102.5	131.8	102.8
31.12.76	60.38	102.8 (HIGH)	132.4	103.6
30.10.78	62.44	92.0 (LOW)	150.9	161.0 (HIGH)
31.12.79	68.46	98.9	155.5 (HIGH)	122.7
3.4.80	71.95	104.6 (HIGH)	149.4	117.8
5.11.80	80.65	100.1 (LOW)	147.8	140.9
28.1.81	83.63 (HIGH)	102.4	145.5	148.2
3.8.81	73.41	117.4 (HIGH)	145.1 (LOW)	133.0 (LOW)
24.9.81	69.13 (LOW)	113.2	150.0	140.7
31.12.81	72.93	111.8 (LOW)	153.1	145.6 (HIGH)
14.6.82	74.09	127.2 (HIGH)	158.8 (HIGH)	134.2

Source: Economic Analysis Section, National Westminster Bank

also places the oil majors under strong cash flow fluctuations giving a very uneven pattern of investment directed as much by the sudden availability of cash as by the commercial response to investment opportunities.

In terms of long-term investor confidence, the experience of the last decade therefore represents a fundamental break with the past. The present economic climate is bleak and the energy investment prospect gloomy. There are, however, lessons to be learned. One of the most important is that we are still trying to make long-term energy investment arrangements and to finance them using the techniques of a previous era—we have hardly begun to cope with the prospect of continuing uncertainty. We are, I suspect, trying to build tomorrow's highways with yesterday's horses. At a technical level at least, therefore, there is a major opportunity to clear the ground and to begin to re-think the stereotypes of company, government and banking finance contracts in the international energy industry.

COPING WITH UNCERTAINTY

An assumption that fundamental uncertainty may persist for a decade does at least tend to eliminate the 'wait until things get better' school of thought, which simply inhibits all decision-making until (probably) it is too late. Rather we need to examine the current contracting techniques, many of which are borrowed from the traditional contracts of the oil multinationals, to see if they are appropriate to new large-scale energy projects. The answer, I suspect, is often in the negative.

The first point is that the structure of international trade in oil is no longer what it was. Government-to-government deals using state corporations or their agents began through the seventies to cover a significant share of oil trade. Although straight government-to-government deals have faded, the role of governments as sellers dominates the oil market. Much of the old vertical integration of the oil multinationals has therefore been broken or is at risk. Crude export contracts have shrunk from a 20–50 year expectation based on equity production rights to a point where three-month break clauses are often the norm and anything over a year is considered long-term. There are many new intermediaries and increasing linkage between crude oil, products, gas, coal and petrochemicals, not only in price terms but in conditionality of joint lifting, processing and marketing. Traditional oil industry thinking which was long-term orientated tends to express itself now in short-term trading terms. But the long-term stable bilateral relationships required in non-oil energy investment require a new flexible framework to establish long-term mutuality of interest.

The second point is that the rapidly changing structure of the oil industry no longer provides a stable norm for the non-oil energy industries. Given the whitening of the oil barrel and the impact of alternative fuels no-one can be sure whether light fuel oil, naphtha or any other product will stand in the same price relationship to other oil products and forms of energy in five or ten years time. The new contracts simply have to cope with these uncertainties.

The third point is that when an oil company invests in a project it generally seeks a minimum commercial rate of return and also the chance of making a high rate of return. The former is calculated by the conventional mechanisms of discounted cash flow in terms which can be readily understood by long-term planners, taxation authorities and engineers, for most of whom the latter objective is an incalculable mystery of the business world. In a regular business cycle in a market economy, the energy industries can usually survive the trough: the dogs bark but the caravan moves on. However when an entire market such as the oil market is subject at intervals to more or less overnight doubling or even quadrupling of its prices, to threats to a large part of its supply and general mayhem, the windfall profits and huge cash flow generated capriciously by company stock positions diminish the relevance of a stable rate of return. The fourth and final point is whether marginal pricing of energy for the consumer dictated largely by the cost of imported oil might often impose quite unacceptable burdens on a national economy. Extreme exposure of certain key industries to outside competition and the inelasticity of demand in many parts of the economic infrastructure may point, for example, to further price decoupling of domestic-produced energy. In global terms, this fragmentation may provide a route to renewed investment in nuclear and other technologies which offer, ultimately, abundant, cheap energy, and greater security of supply.

NEW APPROACHES TO CONTRACTING

Areas where attention needs to be focussed in long-term international contracting include:

1. Review Clauses

Continuing uncertainty implies that arbitrary formulae covering currency, inflation, oil prices, tax or economic activity clauses are unlikely to survive the 15-25 year term of a large international development or supply contract. Parameters and tolerances therefore have to be defined to trigger review or, as may prove more convenient for administrative reasons, standard reviews at regular intervals (e.g. 3 years) may need to be introduced.

2. Penalty Clauses

Any system of regular review will require increased deterrents to breaking the contract. The preservation of mutuality of interest throughout the life of the contract will only be achieved if there are much stiffer penalties for withdrawal which are specified and agreed at the outset.

3. Terms of Supply Clauses

Terms of minimum offtake, interruptions to supply, facilities for taking additional supply with appropriate premia and discounts all need re-thinking in the light of an assumption of continuing uncertainty.

4. Inflation Escalation Clauses

The conventional use of Wholesale Price Indices as a proxy for local industrial costs/purchasing power may need considerable elaboration, particularly where industrial costs diverge sharply for market or government policy reasons from domestic costs. The sometimes poor correlation of the WPI with other price and cost indices emphasizes the need for a continuing review of indices used at present.

5. Oil Price Escalation Clauses

The use of one or two oil product prices as a numeraire has given way to the use of baskets of various crudes and products. No norm has been established as to what basket should be used and this is unlikely as a steady whitening of the barrel and changing structure of the oil industry make it impossible to predict the relative values of oil products, their relationship to crude and their relationship to alternative fuels. As reported in recent oil and gas contracts, the mix of components in the basket itself is likely to be subject to review either at fixed intervals or if it moves outside pre-agreed parameters.

6. Currency Clauses

Bilateral international agreements denominated in the currency of the producer or consumer have shown some of the most bizarre distortions. Hence the efforts of OPEC to formulate SDR denominated agreements and other currency baskets whenever movements in the dollar rate have worked against the interests of the oil producers. The selection of a dollar basis, based on general principles in global oil trading, may have little relevance in, say, a 20 year intra-European gas contract, where each party is trying to tie down its commitment in national budgeting, local currency terms. There is scope for devising safeguards where it can be demonstrated that fluctuations in the bilateral exchange rate derived from unilateral government action or from international commitments concerning official intervention in the exchange markets (e.g. mechanisms such as the European currency snake). Community deals expressed in European Community units of account or international agency financing expressed in SDRs or BIS gold francs may also prove attractive for reasons of achieving political neutrality, and as an insurance against costly fluctuations in exchange rates.

7. Payments Clauses

The need for protection against the imposition of local exchange control and therefore the limitation of convertibility may need considerable elaboration, particularly within the developing world.

8. Financing Arrangements

In conditions of wildly fluctuating interest rates with a mixture of roll-over commercial finance at flexible interest rates and discretionary low cost financing from international agencies and governments, much greater subtlety will be required in drafting the relevant articles. The sequence of major purchasing contracts being used as a security for the financing contracts may be frequently broken where, as in the case of the financing of the USSR gas pipeline, the financing of the project becomes part and parcel of the purchasing deal.

9. Government and International Agency Guarantees

Government backing of large-scale energy contracts is likely to increase. The form of this support is likely to require much closer definition, particularly where, as in the case of the recent French and Italian gas deals with Algeria, the energy project forms part of a much wider trade, investment or financing package.

10. Jurisdiction of National Regulatory Bodies

Given increasing environmental control and government involvement in energy contracting, the territorial jurisdiction of regulatory agencies will require close definition in large contracts. For the United States, the Panhandle-Trunkline hearings before the Federal Regulatory Authorities in November/December 1982 may provide a landmark in the literature of this topic.

NEW ROLES FOR THE FINANCIAL SECTOR

The formulation of finance contracts is essentially a matter for the banks and finance lawyers to get on with. Already that very small group of New York banks from whom much of the innovation in energy finance has come, is turning its attention to the problem. At the industry level their clients are the many US gas and oil independents stimulated by oil and gas price deregulation. It may be that many of the legal and procedural difficulties of the new energy mega-contracts will be ironed out at this micro-level. It is an irony that, on the European side, the terms of the new USSR gas contracts, resisted so vigorously by the United States on security grounds, may well make the breakthrough to more flexible long-term supply contracts in the West in general.

One feature of the increasing emphasis on short-term oil trading is the worldwide interest in developing financial mechanisms to hedge against short-term price and supply fluctuation. So far the new London and New York oil futures markets in oil products have been used for small-scale hedging, but the prospect of their extension in scale and range to cover coal, crude oil and a wider range of oil products has also attracted the attention of economic strategists. They have spotted the opportunity thereby to mobilise government and private oil and coal stocks to smooth sharp oil spot-price fluctuations and also to mobilise bank funds to finance those stocks. The refining industry has also begun to realise that a sophisticated futures market for both crude and product would cover their individual price exposure during the refining process and in aggregate ultimately dictate more precisely future levels of demand and capacity utilisation. A world futures market in crude and product would, by its scale, guarantee for example oil income, as budgeted for by a producer government given a planned volume of production, or define for a consumer government up to a year in advance the precise cost in convertible currency of its oil bill. The full-scale integration of global energy and money which this would imply is probably beyond our reach but at least the relationship between oil and money is beginning to attract much more serious attention.

THE SANCTITY OF CONTRACT

In the final analysis, new front-end loaded international energy megaprojects will only be financed by international money and capital markets if there is adequate investor confidence. This has been profoundly shaken by the collapse of a number of major government-backed energy projects and the reneging on the terms of certain major supply contracts where the finance had already been mobilised and the front-end investment made. Moreover, governments are also only just beginning to come to terms with a much wider issue—establishing a consensus on what level of energy imports constitutes a really serious risk to political security. They need to be fairly sure about that before they can turn away cheap imported energy in favour of more costly investment in 'secure' resources and increased energy efficiency. Unless government backing for major international energy projects is forthcoming and its credibility can be reinforced by agreed international arbitration procedures and sanctions, the financial markets will have an uphill struggle for some years to come.

THE MULTINATIONALS

There is one field where there is a genuine international energy market. This lies in the area of human resources—the world market in human energy, skills, experience and enterprise. All major energy projects today require access to the very latest technology, a wide variety of expertise and effective international co-ordination. The free flow of this human resource is therefore essential.

Successive advances in transport and telecommunications have accelerated and reinforced this process. In recent years, the scale of energy exploration and development has increased massively, involving a multiplicity of governments, companies and individuals. But at the heart of the free world system still lie the major corporate organisations which are multinational in character. They, above all others, have the resource and drive to apply the latest appropriate technology wherever it is needed. They operate simultaneously in arctic icefield and equatorial swamp, in the depths of the North Sea and the sands of Arabia, switching their well-trained and experienced workforce and the latest equipment easily between them. In this flexibility lies their immense resilience and vitality.

Each government and economy will, of course, find its own way of protecting its own national interest with its own institutions, licensing, royalties, taxation and other forms of control. Increased communication and comparative study will show up the anomalies. But if governments persist in interfering to the extent of blocking the free international flow of skill and technology or seriously distorting the optimum development of global energy resources, they will only do so at the peril of our long-term global welfare and prosperity.

SECTION 4

OPENING AND CLOSING ADDRESSES
FULL LIST OF PAPERS

Opening Address

*Sir William Hawthorne**

Mr President, Ladies and Gentlemen, it gives me great pleasure to welcome once more this distinguished Association on the occasion of its second meeting at Churchill College. In welcoming you to the College, I also welcome you to the University. To those of you who have not been here before, I should perhaps explain that this University is of great antiquity but moderate size. It has some 8,000 undergraduates about equally divided between the Arts and Sciences—for some reason we regard Economics as an Art. Behind its medieval facade—the University's, not Churchill's—several thousand graduate students, research workers and Faculties advance knowledge in subjects ranging from animal physiology to robotics, from molecular biology to radio astronomy and, of course from economics to energy. All these are on the whole peaceful pursuits. In fact, the last battle in Cambridgeshire was fought in 905, except for the battles in our Faculties of English and Economics. The battle between public and private enterprise has been resolved by an armistice.

In the University, which is largely supported by public money, there are nearly 30 Colleges, privately endowed and self-governing. Churchill College is the national memorial to Sir Winston and has in its 21 years of existence done very well academically. This year five out of the eight Churchill students taking the Part II Economics Tripos got Firsts. Churchill College has something in common with Trinity College, founded by Henry VIII—its Master is chosen by the Crown and not by the Fellows as, for instance, in C. P. Snow's college.

*Master of Churchill College, Cambridge.

It also has something else in common with Trinity. During the first 21 years of Trinity's existence, there was an energy crisis. Between 1550 and 1600, the price of a cheap, convenient and clean fuel—firewood—went through the roof, i.e. it actually increased about five-fold. Fortunately a less clean but plentiful fuel—sea coal—was available to rescue our UK economy. Apart from some fluctuations the last 300 years has seen a downward trend in the real price of coal. No doubt we shall hear today or tomorrow whether this trend is likely to continue.

When you met here two years ago, there was a parallel meeting going on in Venice—and that summit and your meeting were able to come to some conclusions. The last two years have produced some obfuscations in the energy scene which give this meeting several challenges. I am happy to see that in your extensive programme you have a session on 'Responses to Disruption and Crisis', because the fear of breakdown in supply is justifiably greater than that of a gradual rise in price. I also welcome the attempts to do a better job than economists have in the past at price forecasting. If only someone could give some more reliable figures of present value for energy projects maturing in the next twenty years, we engineers would not have to depend so much on our own guesses and hunches.

Once more, may I welcome the Association and many old friends to Churchill and wish you a comfortable and profitable meeting. I note that they are offering you punting on the Cam. Be careful, punters—you may suddenly have to decide whether to hang on or let go of that pole.

Closing Address

*Paul Tempest**

> *2.50 pm Wednesday 30th June 1982*
> The Chairman and members of the British Institute of Energy Economics cordially request the pleasure of the company of all Conference participants in the annual Pilgrimage by Punt towards Grantchester where, in the Orchard, strawberries, honey and other refreshment may be available for tea.†
>
> *[from the Conference Programme]*

Several friends and colleagues here today seem to think that our suggestion, printed in the Conference Programme, to go punting this afternoon was intended as a joke. Not at all. Punting at Oxford and Cambridge is far too serious a matter for joking.

Now *The Times*, last Wednesday, reported us as planning to punt *to* Grantchester. The Programme states clearly *towards* Grantchester.

May I assure you that, in accordance with the Programme, we *will* be stepping into punts for about an hour at ten to three this afternoon. We have a crate of

*Chairman of the 1982 Conference.

†Topics suggested included Upstream Energy (first half) and Downstream Energy (second half) with, in parallel, The Viscosity of the Banks as Scenario Parameters and Profile Anti-Reviabilitisationism.

wine and a basket of cherries and ripe peaches and hope you will be able to join us.

So I should like to close this Conference with a serious forecast and policy guideline concerning *human* energy.

We may think we are going uncertainly against the current towards what we think is our objective. We will *not* get to Grantchester. There will be a lot of talk and some surprises. But what is important is that, on our frail craft for our brief spell, as many of us as possible should enjoy the trip. Thank you.

List of Papers

All the papers are registered in the Archive of the British Institute of Energy Economics, which is held in Chatham House, 9 St James's Square, London SW1. Unless marked by * copies can be ordered at a cost of £4.00 each, including postage. The figure in brackets gives the number of pages of the full text.

M. A. ADELMAN	(MIT) — The Changing Structure of the Oil and Gas Market (7)
F. E. BANKS	(Uppsala) — Soviet Gas and the Western European Energy Shortage (21)
SIR P. BAXENDELL	(Shell) — Forces and Forecasts in the Energy Business (8)
R. BELGRAVE	(BIJEPP) — Oil Crisis Management (5)
E. BERNDT	(MIT) — Energy Demand: The Choice Between Input and Output (17) (with G. CAMPBELL WATKINS)*
T. W. BERRIE	(Ewbanks) — Interactive Load Control and Energy Management Systems (28) (with B. D. MALLALIEU and K.R.D. MYLON)*
L. G. BROOKES	(Bournemouth) — A Model for Energy Price Hikes on the World Economy and the Long-Term Macroeconomic Role of Nuclear Energy (26)

*Not to be photocopied, distributed or quoted without the author's permission.

M. CARPENTIER	(Director-General, E. C. Brussels) — The Energy Problem: The Response of the European Community (12)
M. CONANT	(Conant Associates) — The Strategic and Economic Aspects of Gulf Gas (9)
J. H. CULHANE	(Occidental) — The Changing Economic Climate of the Oil Industry (17)
R. DAFTER	(Financial Times) — Experts Agree to Disagree on Next Oil Shock
T. R. DEALTRY	(RBA) — Associated Gases and their Application in the Middle East Oil Producing States* (15)
R. DEAM	(QMC, London) — Oil and Gas Producer Revenue: A Possible Solution to the Problem of Decreasing Profit (13) (with PROF. M. A. LAUGHTON)*
S. DEVARAJAN	(Harvard) — Drawing Down the Strategic Petroleum Reserve: A Case for Selling Futures Contracts (35) (with G. HUBBARD)
C. K. EBINGER	(Georgetown) — Energy Crises and the Third World—An Opportunity for the West (34)
J. B. EDWARDS	(US Secretary for Energy) — US Energy Policy (8)
N. EVANS	(Cambridge) — Electricity Investment Planning in the UK* (8)
F. FESHARAKI	(Honolulu) — Hydrocarbon Processing in OPEC Countries; Excess Capacities and Readjustment Pains in the World Refining Industry (62) (with D. T. ISAAK)
J. T. FRASER	(Tennessee Gas) — Prospects for US National Gas Imports Post Deregulation: A Supply-Demand Analysis (63)
M. FRITZ	(UNESCO) — Problems in Implementing Alternative Energies in the Developing World (24)
D. GALLAGHER	(University of New South Wales) — Natural Resource Scarcity: The Case of Australian Coal (36) (with G. D. McCOLL)
W. GREAVES	(NEDC) — Prospects for Petroleum Futures Markets (23)
T. GREENING	(1st National Bank of Chicago) — Finding and Financing the Next Trillion Barrels (22)
W. HAWTHORNE	(Churchill College, Cambridge) — Conference Introduction* (2)
J. E. HARTSHORN	Remarks on the Final Panel Discussion* (2)
L. HOFFMAN	(Regensburg) — The Impact of Rising Oil Prices on Oil Importing Developing Countries and the Scope for Adjustment (51) (with L. JARASS)
C. HOPE	(Cavendish Laboratory, Cambridge) — Adaptation

D. MOUNTAIN	(Ontario Hydro) — The Influence of Higher Energy Prices on Canadian Agricultural Productivity (25) (with M. W. L. CHAN of McMaster University, Hamilton)*
R. NETSCHERT	(NERA) — Interfuel Competition and Energy Policy in the OPEC Area (24)
P. R. ODELL	(Erasmus University, Rotterdam) — Towards a System of World Oil Regions (11)
J. F. O'LEARY	(J. F. O'Leary Associates) — Price-Reactive versus Price-Active Energy Policies (22)
P. O'SULLIVAN	(University of Wales) — Energy Conservation Policy in the Building Sector (13) (with R. WENSLEY of the London Business School)
R. K. PACHAURI	(West Virginia University) — Changing Markets for Coal in the Developing Countries (28) (with W. LABYS)
E. PARR-JOHNSTON	(Shell Canada) — Mega Project Energy Development as the Basis for Industrial Development: Some Potential Problems (21)
J. PLUMMER	(President, QED, Palo Alto) — The Current International Deadlock over Stimulating Oil Exploration in Less Developed Countries (7)
M. V. POSNER	(Social Science Research Council) — After Dinner Address*
I. C. PRICE	(Sir William Halcrow) — A Scenario for UK Electricity Supply and Demand in the Year 2012 (12)
S. A. RAVID	(University of Haifa) — A Cost Benefit Analysis of Bio-Gas Production — The Israeli Experience (34) (with J. HADAR)*
K. H. RISING	(Battelle, Pacific Northwest Laboratories) — A Decision-Making Model for the Recovery of Useful Material Resources from Waste* (14) (with G. A. JENSEN and V. F. FITZPATRICK)
C. ROBINSON	(University of Surrey) — The Future Oil Price (20)
J. ROEBER	(Joe Roeber Associates) — The Role of Government in the Oil Industry*
H. D. SAUNDERS	(Tosco Corporation) — Optimal Oil Producer Behaviour Considering Macrofeedbacks* (29)
A. F. G. SCANLAN	(BP) — The Impact of Communist Bloc Energy Supply and Demand (17)
L. SCHIPPER	(University of California, Berkeley) — How Much Energy Have We Saved—Residential Energy Use in the OECD 1960-2000* (38) (with A. KETOFF)
S. H. SCHURR	(Electric Power Research Institute) — Energy Efficiency and Productive Efficiency: Some Thoughts Based on American Experience (20)
M. SIDDAYAO	(The East-West Resource Systems Institute) — Oil

	Importing Developing Countries: Capital Formation, Balance of Payments Constraints and Manufacturing Progress (29)
J. STANLEY-MILLER	(Dept. of Energy, Washington) — A Short-Term World Oil Price Market Model* (with J. M. CORLISS)
J. L. SWEENEY	(Stanford University — World Oil* (2)
J. E. TREAT	(New York Mercantile Exchange) — Energy Futures Markets (4)
N. TRIMBLE	(British Gas Corporation) — Gas Prices and Exploration (3)
R. S. UHLER	(The University of British Columbia) — Prices of Oil and Natural Gas Reserves in Canada (34)
P. VERLEGER	(Booz Allen & Hamilton) — Rational Strategies for Meeting Oil Market Disruption (9)
R. WEINER	(University of Harvard) — The Sub-Trigger Crisis—An Economic Analysis of Flexible Stock Policies* (34) (with G. HUBBARD)
K. WIGLEY	(UK Department of Energy) — Methods for Projecting UK Energy Demand Used in the Department of Energy (24)
J. WILKINSON	(Sun Oil) — The World Oil Market: A Short-Term Perspective (11)

SECTION 5

ABSTRACTS

PART II

The World Oil Market: A Short-Term Perspective

*Jack W. Wilkinson**

This paper summarizes efforts to date concerning the linkages between short-term world macroeconomic activity and the world primary energy market, with particular emphasis on world oil supply, demand, and price.

For this effort, small-scale macroeconomic models were developed for each of the seven major industrial countries—Canada, France, Italy, Japan, United Kingdom, United States, and West Germany—plus an aggregate for the rest-of-the-free-world. These models include: expenditure, supply, price, labor, finance, and exchange rate sectors. The individual models are linked through a simple trade matrix.

*Chief Economist, Sun Co. Inc.

The Effect of Market Structure on Patterns of International Coal Trade

*David S. Abbey and Charles D. Kolstad**

The purpose of this paper is to explore the significance and effect on patterns of coal trade of three types of deviations from the simple competitive model. On the producer side duopoly (South Africa, Australia) with a competitive fringe (US, Canada, Poland, Colombia) is examined; on the consumption side, regional monopsonists (Japan). Finally, importer security of supply through imports portfolio selection is considered. A simple model of coal trade is used to examine the extent to which these factors can explain existing and anticipated trade patterns.

*Los Alamos National Laboratory.

Electricity Investment Planning in the UK

*N. Evans**

The familiar technique of finding some least cost expansion programme to satisfy projected demand for electricity in England and Wales has been extended to include an appraisal of past investment decisions. In comparing actual past performance with an idealised optimal path it is concluded that it is not primarily plant over-capacity but adverse mix of generating plant that most increases generating costs.

*Energy Research Group, Cavendish Laboratory, Cambridge UK.

PART III

Energy Demand: The Choice Between Input and Output BTU Measurement‡

E. R. Berndt and G. C. Watkins*†

Energy demand modelling can be critically affected by the point at which energy consumption is measured. Should it be where energy is fed into a furnace or energy system, popularly referred to as 'input' BTUs? Or should it be at the point at which energy is released for end use, popularly called output BTUs? Input BTUs correspond to the amount of purchased energy. Output BTUs correspond to the effective energy output.

The translation from input to output BTUs is usually made by applying a vector of efficiency factors for individual fuels. This paper examines both theoretical and empirical aspects of the 'input-output' BTUs controversy. It lists some demand studies according to preference for input-output BTUs and examines several empirical features of input-output BTU measurement. These aspects are then placed in a broader analytical framework. It argues that energy demand modelling should focus on input rather than output BTUs.

*Professor, Massachusetts Institute of Technology.
†President, Datametrics Limited and University of Calgary.
‡This paper, registered in the BIEE Archive, is in preliminary form and not for citation.

Interactive Load Control and Energy Management

*T. W. Berrie, B. D. Mallalieu and K. R. D. Mylon**

Microelectronics in the Service of Energy Economics

Energy Management Systems are those devices which seek to minimise energy usage and maximise plant usage such that maximum productivity ensues. Though such systems can be human or mechanised, we are considering the latter type in this paper.

Starting from simple, electromagnetic devices, e.g. time switches and relays, etc., energy management equipment quickly utilised the computing facilities which became widely available only a few years ago and initially sophisticated, large, centrally-based systems were developed to monitor and control extensive buildings, campuses or industrial plants, where all loads can be wired back to the main equipment. With the rapid development of micro-electronics these were followed by smaller, microprocessor-based, distributed, intelligent, stand-alone outstations interconnected with centralised supervisory mini or micro computers. The outstations can communicate, when necessary, with the remote central supervisory equipment, which can interrogate the local outstation when data is required or can initiate energy management commands. Outstations can be situated some considerable distance from the central supervisory station and from each other, and make use of direct tie-line or auto dial-up PTT links using modems or use the electricity distribution system (PLC) for communication. This development introduced the potential for an extensive distributed network of intelligent, stand-alone systems interconnected via hierarchical, communication facilities used intermittently. These, in turn, gave birth to 'two-way' interactive control, whereby not only can the outstations be programmed locally by the operator to monitor and control independently their own local operations, but they can also react to the overall power system and consumer load demand changes. They can, therefore, interact with each other to maintain system balance.

*Ewbank and Partners Limited.

Home Energy Use in Nine Countries: How Much have we Saved?

*Lee Schipper and Andrea Ketoff**

In 1979 the Lawrence Berkeley Laboratory (LBL) was asked by the US Energy Information Administration of the Department of Energy (DOE) to gather demographic, economic, and energy data for the residential sectors of a variety of OECD countries. It was discovered that neither the OECD nor most of its member countries kept such statistics for the residential sector in one place.

The countries covered include Canada, Denmark, France, W. Germany, Italy, Japan, Sweden, the United Kingdom, and the United States. The data collected included information about incomes and expenditures, size of households and dwelling characteristics, household equipment, and actual consumption of fuels, often by purpose and dwelling type. The output of the study includes a data base covering energy use from 1960 through 1980 and several analyses of comparative consumption patterns and their evolution. Finally, econometric analysis of the energy consumption data is underway.

An analysis of the structural and energy data leads to five basic findings:

1. Increases in equipment ownership (central heat, running hot water, and major appliances), driven by higher incomes, were responsible for the great increases (4-8% per annum) in residential energy use through 1973 and the sustained growth rates through the late 1970s, although after 1973 the growth rates slowed considerably relative to income growth.
2. Great variations in ownership, lifestyle, and energy intensity occur even at roughly similar income levels. Differences in prices, climate, housing stocks, heating, bathing and cooking habits, and engineering efficiency account for these variations, but in differing proportions.
3. Increased energy prices since 1972 have induced reductions in energy intensities, and, more recently, caused shifts away from oil heating systems where alternatives have existed. Heating energy use has been reduced considerably in centrally-heated homes since 1972.
4. Information programs have probably played an important role in informing consumers of options for conserving energy. Energy conservation programs have affected investments in energy saving equipment

that may not have been undertaken. Still, the effects of these programs on total energy saved appear to be less than the short-term actions taken by consumers alone.

5. When all factors are counted there are still significant differences in efficiency of heating and appliance systems among countries. The best techniques used in one country may allow significant energy savings through their introduction elsewhere.

*Lawrence Berkeley Laboratory.

Energy Conservation Policy in the Building Sector

Professor Pat O'Sullivan and Dr Robin Wensley†*

This paper‡ addresses the problems of policy on the demand side, with particular reference to the building sector, although it is argued that the policy problems are often similar in other sectors. The basic policy mechanism on the demand side in the UK, as in a number of other countries, is the use of an appropriate fuel pricing policy. Although there is evidence that such a policy works in that significant demand changes can be observed as a result of price changes it is also true that on its own such a policy takes a long time to have an effect. The effect can also be substantially distorted by short term fluctuations and information problems for fuel users.

There are three reasons for concern with demand policy measures over and above use of the price mechanism:

(a) high political costs which can accrue in that the adjustment process can result in substantial consumer frustration;
(b) high welfare costs in that although on average consumers do adjust it is often most costly and difficult for those who are in disadvantaged groups;
(c) the need for better demand estimation in the context of supply policy. In a number of areas it is apparent that the lack of adequate demand response information creates considerable estimation problems.

Even in relatively buoyant years new building only accounts for 1–3% per year of the building stock. Directly similar improvements in the case of existing buildings can only be expected in a limited number of up-market properties. To influence energy use in the remainder it is necessary to consider the evidence on actual energy use behaviour and how it varies. Such evidence suggests that generalised attitudes have little impact and that householders are often unable to respond effectively because of variable expectations about future energy prices, lack of information on how to respond and, finally, shortage of necessary funds.

From an economic policy perspective there are clear imperfections in the system, related to tax, risk, time horizon and the separation of costs and benefits. As a result actual market responses will be neither as consistent nor as rapid as the ideal demand policy might require. Hence government intervention is required to recognise particular problem areas and target action to resolve such problems. Examples of such action would include on the conservation supply side, the improvement of standards of workmanship and the encouragement of the development of controls based on anticipation rather than response. On the demand side there are clear needs for useful and specific information on, for instance, how to insulate and draughtproof, as well as encouraging the proper location and design of metering and providing specific financial incentives.

*Professor of Architecture, University of Wales.

†London Business School.

‡Most of this paper is included in *Energy Economics in Britain*, published by Graham & Trotman Ltd.

PART IV

World Oil

*J. L. Sweeney**

This study† conducted by a 43 member working group of the sixth Energy Modeling Forum reports results obtained through the application of ten prominent world oil or world energy models to twelve scenarios. These scenarios were designed to bound the range of likely future world oil market outcomes. The conclusions relate to oil market trends, impacts of policies on oil prices, security of oil supplies, impacts of policies on oil security problems, use of the oil import premium in policy-making, the transition to oil substitutes, and the state-of-the-art of world oil modeling.

The analysis suggests that the key issues are how rapidly inflation-adjusted prices will rise during the next several decades and the duration and magnitude of the short-run price decline. The model-based projections suggest a soft oil market in the first half of the decade unless another supply disruption occurs, but suggest that by 1990, inflation-adjusted prices can be expected to exceed current high levels.

Policies which reduce oil import demand, whether implemented by nations acting alone or collectively, can significantly reduce oil price growth. By 1990 we can expect world oil prices to be reduced by between 90c and $2.40 per barrel (1981 constant dollars) for each 1 million barrels per day reduction in oil import demand. Since the US alone or the OECD nations together import large quantities of oil, import reductions which moderate oil price increases can yield significant economic benefits to these countries. The benefits, when expressed on a per barrel basis, are referred to as the market power component of the import premium. The estimated median value of the market power component is about $8.00 per barrel above the world oil price.

To examine the implications of oil supply disruptions we applied the models to a hypothetical reduction in OPEC's production capacity. An 8 million barrels per day reduction in OPEC production could cause world oil prices to leap upwards by $40–100 per barrel above the levels occurring at the time of the disruption. Economic impacts of such supply disruptions could be immense!

The analysis of disruptions suggests that price jumps could be moderated significantly if oil stocks were released or excess oil production capacity were used during the disruption: a 1 million barrel per day stockpile release might moderate price jumps by around $20 per barrel. Gradual import reduction programs taken in advance of a disruption might lead to reductions in OPEC production and consequently to increases in excess oil production capacity.

Such programs reduce the economic costs of a disruption. For example, a 5 million barrel per day demand reduction undertaken prior to the hypothetical 10 million barrel per day reduction in oil production capacity reduces the economic cost of the disruption by 30% to 90%.

When expressed on a per-million barrel per day basis, the disruption-related economic benefit of oil import reductions is referred to as the 'security component' of the import premium. Under our overly conservative assumptions, a security component estimate ranges from $1 to $5 per barrel for the US alone and three times these values for the OECD as a whole.

Beyond the next two decades or so, far reaching changes in the oil market may begin to manifest themselves. We examined projections of oil market effects of different assumed unconventional fuel supply levels. Under our assumptions, the world oil price rises well above average cost of oil substitutes in most circumstances. Massive production of these substitutes is required before world oil prices decline to their average production. While these 'backstop' technologies may not place a cap over world oil prices during the next thirty years, generally the greater the quantity of backstop production, the lower the world oil price.

*Stanford University.

†The full text is not available through the British Institute of Energy Economics.

The Future of Crude Oil Prices

*Colin Robinson**

Conclusions

Oil price predictions in the past have not been so strikingly successful that we can say anything with great confidence about how prices may move in the future. The opinion I would venture, however, is that if we try to form a judgment by looking at likely demand and supply influences the most reasonable conclusion we can reach is that expectations will in general be for real oil price increases and that real prices will probably rise in the 1980s and the early 1990s. The plausible scenarios seem to me to be those in which real oil prices fluctuate a good deal, but about a moderately rising trend. It appears likely that occasional supply shortages (actual or perceived) and spurts of demand about a slightly increasing world demand for oil are likely periodically to drive up real prices in steps; though there will no doubt subsequently be some downward drift caused by inflation and discounting by producers, it would be surprising if the price floor and the price ceiling were not rising. In other words we may in the 1980s and early 1990s see several small-scale repetitions of the events of recent years. The average rate of increase of crude prices in the rest of the 1980s and in the 1990s is a matter about which, in all honesty, we can only speculate: I would take as a working assumption real annual average increases in the range of 1 to 5 per cent per annum, though the upper end of the range does not at present seem very probable.

Therefore, although I have since 1973 argued that big oil price rises in the 1970s would turn out to be a temporary phenomenon (since the market would adjust and damp down the increases), I do not think we have yet reached the stage when we can reasonably anticipate falling or even constant real prices. In the long run we cannot expect oil prices to continue rising relative to the price level in general—if they were doing so the supply and demand shift would be so large and the behavioural changes so great that the real price rise would cease. But the time lags in the system are sufficiently lengthy that the long run in this context may mean the late 1990s and early next century rather than the next ten years.

*Professor of Economics, University of Surrey.

Gas Prices and Exploration

*Niall Trimble**

Over the last year or so there has been a great deal of comment about the prices paid for North Sea gas by the British Gas Corporation. A number of commentators have suggested that the reserve potential of the North Sea is very large but the low prices offered by British Gas have destroyed the incentive to explore. As a result, they claim that exploration has fallen to fairly low levels and the high reserve potential has not been reflected in the discoveries made. This paper examines the position on exploration and reserves to show that this view of matters can be very misleading and that exploration levels have been influenced by a range of factors, some of which may be more important than price.

*Energy Policy Department, British Gas Corporation.

Methods for Projecting UK Energy Demands used in the Department of Energy

*K. J. Wigley and K. Vernon**

The first part of the paper† describes the approach used for preparing energy projections in the Department of Energy. A brief outline of the overall model is provided and the variations in main assumptions employed to produce alternative projections are indicated. The second part of the paper provides greater detail on the energy demand models developed for the domestic and industrial sectors (excluding iron and steel). Discussion is provided on estimation, stability of coefficients, goodness of fit, price and activity responses and dynamic properties. Some alternative and possible future developments on demand modelling are indicated.

*UK Department of Energy.

†This paper is included in full in *Energy Economics in Britain* published by Graham & Trotman Ltd.

Experts Agree to Disagree on Next Oil Shock

*Ray Dafter**

Invariably, energy economists hedge their bets by admitting all their forecasts could be overturned by shocks and crises. No one knows what will trigger the next energy crisis; that is the point about shocks. But the Association has conducted an exercise which at least points to some of the main fears troubling industry analysts.

In a unique poll, conducted at the end of the conference, the energy economists from around the world—but predominantly from Europe and the US—were asked to state what they considered be a 'likely' shock to the international oil market. They were then asked to think of a 'less likely' crisis.

The psychology behind the two questions is interesting in itself. It was assumed that answers to the 'likely' shock question would be influenced by a consensus view of worldwide tensions and that the second question would elicit the more personal concerns of the economists.

In the end, there was little doubt about the likely seat of the next crisis: the Middle East, a region which accounts for 53.5 per cent of the world's proven oil reserves and over 27 per cent of total oil production.

What was striking was that a revolution in Saudi Arabia was considered the most likely event to cause the next major 'oil shock'.

The threat of a Middle East war and possible Soviet aggression in the Gulf were also cited among the concerns. Those who were specific about such issues saw the Arab-Israeli confrontation as the most likely cause of a war, and Iran as the most likely target of a Soviet invasion.

Outside the Middle East the issue that seems to be worrying energy forecasters the most is the vulnerability of nuclear power. Over 10 per cent of the respondents thought there was a possibility of a nuclear accident leading either to a curtailment of the world's atomic energy programme or, even worse, a complete shut-down of the industry. Such a move could have a significant impact given that nuclear power now accounts for about 5 per cent of energy consumption in Western Europe and about 6 per cent in Japan.

**Financial Times.*

PART V

Interfuel Competition and Energy Policy in the OPEC Era

*Bruce C. Netschert**

Conclusions

The obvious prescription for energy policy in the 1980s is: permit the price system to work to the fullest as the best means of coping with the continued exercise of OPEC power. This can be done in three ways, the first of which is to remove institutional barriers. Eliminate subsidies wherever possible, so that interfuel competition can work through the correct price signals. Use suasion and the hortative power of government to begin the creation of the infrastructure for large-scale methanol use. (One of the institutional obstacles to that achievement is the current industry structure on the supply side. The oil industry, with its investment in refineries, has no particular interest in pursuing the development of a fuel that bypasses the refinery stage. It is no coincidence that all the synthetic liquid fuel options—shale oil, tar sands and coal liquefaction—yield refinery feedstock. The role of government is thus to make the public aware of the methanol option and thus create the basis for a methanol automotive fuel market.) Keep up the pressure to use marginal cost as the pricing criteria for electric and gas utilities. Insist that the internationalization of environmental costs meet the cost-benefit test, so that prices are not distorted in the opposite direction.

The second way is to directly sap OPEC's strength. Part of that strength is the ability to raise prices during partial supply interruptions, as at the time of the Iranian Revolution. The inability to do it at the outbreak of the Iran-Iraq war was a direct consequence of the size of worldwide stocks. The International Energy Agency agreement on proper stock levels and the accumulation of government stocks such as the Strategic Petroleum Reserve of the United States are both correct policy. The cost of carrying such excess inventory is appropriately payable by governments as the insurance premium against a second embargo. (Given the changed relationship between the international oil companies and the producing countries, a second embargo would be far better enforced than the first one.) OPEC's strength can also be attacked directly by the imposition of tariffs or excise taxes to capture the rents, as has been urged by several economists.

Finally, there is benign neglect (to use a well-known phrase). Given time, as we have seen, the price system will work. Doing nothing is better than doing many if not most of the things attempted by the United States government in the second half of the 1970s. On the other hand, the assumption at that time was that the price system would not work. Now that we know it does, there is hope that government policy will be used to complement and facilitate the working of the system rather than to thwart it.

*Vice President, National Economic Research Associates, Inc.

A Simple Decision Making Model for the Recovery of Useful Material Resources from Waste

*K. H. Rising, G. A. Jensen and V. F. FitzPatrick**

In the United States, many of the material resources necessary for energy production are imported. Strategic stockpiling of these resources has been a well-known method for reducing the economic impact of supply interruption in case of emergency. Another viable option, often overlooked, is the recovery of valuable materials and recycle of useful products from wastes generated in the energy production and other chemical processes. The technical feasibility for recovery and recycle, including decontamination of nuclear related materials, has been proven and demonstrated. The economic feasibility would depend on both the intrinsic and strategic values of the material, the net energy saved from reusing rather than producing the material, and other factors that may influence the incentive for recovery and recycle. In this paper, a rule-of-thumb model is developed to quantify the strategic value, evaluate the incentive, and determine the economic feasibility of recovery and recycle. The simplicity of this model allows quick investment decision making instead of the more laborious approach which requires greater detail and many subjective judgments involving the probability and consequence of a foreign supply interruption.

*Battelle Pacific North West Laboratories.

A Cost Benefit Analysis of Bio-Gas Production: The Israeli Experience

*A Ravid and J. Hadar**

The paper presents a model incorporating technological and economic data to obtain a cost benefit analysis of a bio-gas production plant. (The specific example discussed is that of a facility planned by Kibbutz Beit-Hashita.)

After an optimal plant size is obtained, a profit function is calculated and the cost benefit analysis is performed. It is found that bio-gas, given existing technology, is a viable alternative for replacing LPG, but is more expensive than heavier fuels. This cost disadvantage, however, may be offset by environmental advantages of burning gas.

The results obtained for the specific plant analyzed, are extended, and the potential for bio-gas production for the entire State of Israel is examined. It is found that bio-gas could replace 1-2% of industrial fuel consumption in the State.

*Haifa University.

PART VI

Crude Oil and Natural Gas Reserves Prices in Canada

*Russell S. Uhler**

The purpose of this paper is to calculate time series of oil and gas reserves prices for the Province of Alberta, Canada. Such series are of interest in their own right but they also provide price data upon which to base the estimates of price response in a variety of reserves supply models. Although the calculation of these prices makes use of standard economic concepts in asset valuation the application of these concepts to the valuation of oil and gas reserves requires some explanation because of the nature of the asset in question. Particularly important is determining the optimal extraction rate of a reservoir. This issue is considered at some length in the second section of the paper where it is shown that natural reservoir characteristics are likely to dominate economic factors in determining the optimal extraction rate from oil and gas reservoirs.

Another way of determining the price of oil and gas reserves is to use the finding cost approach. Under competitive conditions and perfect certainty firms should discover reserves at a rate up to that amount where their marginal finding cost equals their price. Thus, if one knows the finding cost then the price is determined. In this paper the results on oil finding costs in Uhler (1979) with the oil reserves prices calculated in the way indicated in this paper are compared and reasons offered for the rather substantial differences observed.

A focal point of this paper is on the impact of the tax treatment of the oil and gas industry in Alberta on reserves prices and thus on exploration incentives. The impact of the National Energy Program is also discussed.

*Department of Economics, The University of British Columbia, Vancouver.

The Effects of Taxation on Petroleum Exploitation: A Comparative Study†

*Alexander G. Kemp and David Rose**

Abstract of Conclusions

In this paper fifteen fiscal regimes applied to oil exploitation have been examined to discover their effects. The systems cover the UK, Norway, Australia, Indonesia, Nigeria, Egypt, Malaysia, Papua New Guinea, US OCS (four variants), Alaska (two variants) and Texas.

The results indicate great variations in the level of tax burdens as well as in their structures. Some systems are progressive, others proportional and others regressive across fields of differing profitability. In money-of-the-day terms the Australian system is the most lenient and is broadly proportional. The Texas system is also comparatively lenient and proportional. The US OCS, Indonesian, Egyptian and Nigerian systems are all broadly proportional at increasing levels of severity. The Alaskan systems and the Malaysian one are regressive. The British and Papua New Guinea systems are generally progressive as is the Norwegian one to a much less extent.

The implications of these findings are that the fiscal systems are unsympathetic to the needs of risk-averse investors: the real tax burdens increase when operating circumstances become worse. It is all the more unfortunate that this finding operates most strongly on the least profitable fields. The root of the fiscal trouble is the use of taxes based on production which are not very responsive to deteriorations in operating conditions.

The UK system produced the lowest risk (as defined) closely followed by Norway and Papua New Guinea. Systems with large non-profit related elements in their tax system such as Malaysia and Alaska with sliding-scale royalty were found to cause more risks to investors. It was interesting to find that the country with the most severe tax system—Malaysia—caused the greatest risks to investors.

*University of Aberdeen.

†The conclusions of this very detailed study are published here. The full 64-page text with charts and tables is available in the BIEE Archive.

Finding and Financing the Next Trillion Barrels: A Review

*Timothy Greening**

Because the future supply of non-OPEC oil depends on geological opportunities and economic constraints, both are considered in this paper. First, recent trends in exploration are reviewed; they show disparities between the apparent pattern of geological and economic opportunity and the actual pattern of exploration activity. These disparities can be explained in part by economic and political barriers to investment. The second part of the paper presents alternative views of the geological possibilities. The consensus among geologists and statisticians is that ultimate recovery of conventional oil will be about 2 trillion barrels, of which half has already been produced or added to proved reserves. In the final section an attempt is made at a comprehensive listing of investment barriers, particularly the non-tax barriers. Non-tax barriers have not been documented as thoroughly as the problem of prohibitively high tax rates but are often equally important in determining the economic attractiveness of investment in individual countries.

*Energy Economist, The First National Bank of Chicago.

PART VIII

Mega Project Energy Development—as the Basis for Industrial Development—some Potential Problems

*E. Parr-Johnston**

This paper focuses on the problems inherent in a national development strategy built on mega projects and argues that many of the difficulties inherent in such a strategy were identifiable in advance, e.g. before the recent demise of the Alsands project or the further delay of the Alaska Gas Pipeline. The analysis examines the strategy and concludes that policy at the federal level must be balanced, flexible, and comprehensive rather than speculative and narrowly focused so that the natural resource sector's important role in national economic progress is put into a proper perspective rather than escalated to a risky single engine status with inadequate consideration of the overall policy environment's impact on the sector and the projects.

*Manager, Macro-Environment Corporate Strategies, Shell Canada Ltd, Toronto.

PART IX

Internal and External Effects of an Oil Import Tariff

*Laurence R. Jacobson and Ralph Tryon**

While many analyses of an oil import tariff consider the effect of the tariff on the level of prices and activity in the domestic economy, they often ignore the external repercussions on the current account balance, exchange rates, and secondary effects on other countries. By using a multi-country model with endogenous exchange rates and trade flows, we are able to examine internal and external channels of transmission of an oil import tariff which may not exist in other models.

With a unilateral US tariff, the fall in petroleum consumption resulting from the oil import tariff leads to a current account improvement and hence to an exchange rate appreciation. The currency appreciation lowers import prices, but also results in a reduction in exports and thus GNP. The domestic oil price increase has a positive direct effect on the price level, but this is partly offset by the effect of appreciation. If the government retains the tax revenue from the tariff (rather than redistributing it through tax cuts or transfers), the domestic economy is depressed further. In fact, a reduction in nominal GNP which reduces income tax revenues substantially offsets the revenue gains from the tariff. If the oil tariff is imposed by several countries simultaneously, the improvement in the US current account and dollar appreciation are reduced and the decline in real US GNP is lessened.

In summary, the most surprising, quantitative results concern the importance of the fiscal policy assumption in determining the effects of a tariff, the large impact on bilateral exchange rates caused by tariff imposition, and the substantial impact on economic activity through both internal and external channels. Since exchange rate changes in particular have proven extremely difficult to predict, the quantitative results should be regarded with a reasonable amount of skepticism. Also, having no endogenous model of world oil price determination, we have used an extreme assumption (no oil price change) which maximizes the impact on the domestic economy. However, we hope we have demonstrated that an oil tariff may have significant repercussions through external channels which should be taken into account.

*Both authors are on the staff of the Federal Reserve Board (International Finance Division) in Washington. Details of the Multi-Country model used and methodology are included in the full paper registered in the BIEE Archive.

Oil Crisis Management

*Robert Belgrave**

The author of this paper has recently been engaged in an analysis of the events of the two oil crises, following the Iranian Revolution in 1978/79 and the outbreak of the Iraq/Iran war in 1980. This work was done in collaboration with Mr Daniel Badger, to whom the relevant statistics and records of the International Energy Agency were made available. It was published in

May 1982 under the title 'What went right in 1980?' by PSI for the British Institutes' Joint Energy Policy Programme.

This paper considers the need for preparations to deal with any future oil supply crisis of similar magnitude, and the best way to avoid rapid escalation of prices, such as took place in 1979.

*British Institutes' Joint Energy Policy Programme/Royal Institute of International Affairs.

The 'Sub-Trigger Crisis': An Economic Analysis of Flexible Stock Policies

*R. Glenn Hubbard and Robert Weiner**

In this paper, we compare the economic effects of various proposed 'sub-trigger' stock policies. The effects of unilateral American action on other buyers in the world oil market (as in the case of the 1979 middle distillate entitlement) is well known. *We turn the question around by analyzing the effects of (non-US) OECD stock policies on the American economy*, utilizing a macroeconomic model linked to a model of the world oil market. The 'flexible stock policies' analyzed are as follows:

1. Countries rely on the market until the IEA agreement is triggered.
2. USA relies on the market, other countries agree to draw down.
3. USA relies on the market, other countries build up stocks.
4. Unilateral US SPR draw down, no international coordination.
5. All countries participate in 'sub-trigger' draw down agreement.

We evaluate these policies by simulating a 'sub-trigger' disruption which grows over time, as happened in 1978-79, when Iranian production steadily diminished, and Saudi Arabia first increased, then curtailed output. Eventually, the disruption is large enough to trigger the IEA mechanism. We show that cooperation makes a significant difference for US economic performance, whether or not the government draws down the SPR, and that sub-trigger action can be preferable to waiting until the disruption becomes a conflagration. We conclude with recommendations for US policy.

*Harvard University.

Drawing-down the Strategic Petroleum Reserve: The Case for Selling Futures Contracts

Shantayanan Devarajan and R. Glenn Hubbard†*

Conclusions

In this paper, we have attempted to make a case for drawing-down the strategic petroleum reserve by selling futures contracts. The motivation behind this idea is the observation that, during oil supply disruptions, private inventories act in a manner to exacerbate the sharp rise in oil prices. We showed how sales of SPR futures, unlike other policy responses, dampen inventory demand by affecting expectations of future oil prices. Using a two-period model we identified conditions under which the sale of an equivalent amount of SPR oil in the spot and in the futures market would lead to a more favorable sequence of spot prices in the latter case. Our simulation results, based on a model of the world oil market and the US economy, showed that futures sales (i) achieved much of the price-reducing benefits in the early stages of a disruption and (ii) led to a lower price trajectory overall when compared with spot market sales. Finally, we identified other possible advantages of SPR futures, which center mainly around the fact that futures contracts postpone, and in some cases avoid, the physical withdrawal of oil from the reserve.

Our results represent a first-pass at analyzing a new policy option for responding to a disruption in the oil market. Both our theoretical and empirical analyses can be extended to incorporate more realistic features of the world oil market. Yet, we suspect the qualitative nature of our results will remain unchanged: selling futures contracts in SPR oil can be at least as effective in dampening spot market prices as a direct sale in the spot market.

On the other side of modelling and estimating these benefits, there lies the question of how such a scheme may be implemented. As there currently exists only a very thin futures market in crude oil, and as attempts to set up such a market in the past have met with limited success, the procedure by which the government could sell SPR futures contracts deserves considerable attention. Although we believe sales of SPR futures during a disruption are feasible, and in many ways easier to effect than spot market sales, we also feel that the success of such a policy depends on the amount of thought given to its implementation before the interruption occurs.

*John F. Kennedy School of Government, Harvard University.
†Department of Economics, Harvard University.

The New York Market in Petroleum Futures

*John Elting Treat**

This paper explains how commodity trading works, the purpose of commodity markets, and gives an overview of worldwide commodities trading. It then examines the impact of futures trading on oil markets, and explains the New York markets in petroleum futures.

A well developed futures market can be used as a price discovery tool promoting increased competition in the petroleum product markets. Futures markets prices are very reliable since trades in the futures markets are based on cash commitments. Moreover, these prices are widely disseminated and available to everyone regardless of whether they are involved in the futures market.

By using futures contracts to hedge future purchases and sales, prices can be 'locked-in' six months or more in advance of the actual transaction. There are a myriad of uses for such a hedge tool. For example, in the US capital, the Washington Metropolitan Area Transit Authority, which operates the area's buses and subways, is in effect utilizing the futures market to fix their fuel acquisition costs over a 12-month period. The agency has thus minimized the likelihood that they will need to request additional revenues in the middle of the budget cycle or need to impose sudden, unpopular fare increases. Similar approaches are now being developed for other major purchasers, including the US Dept. of Defense.

In the future, it is possible that oil producers might use the crude oil market, which will be opened on the New York Mercantile Exchange, to guarantee a certain level of petroleum revenues, thereby facilitating government budget planning. Representatives of Venezuela, Ecuador and Brazil have been briefed concerning hedging in the petroleum futures market at NYMEX.

*President, New York Mercantile Exchange.

Petroleum Futures Markets

*Walter Greaves**

This very detailed paper† examines the history of the Petroleum Futures markets in New York and London and the present state of play. It concludes:

'The previous emphasis on the integration of crude oil supply, refining and marketing is being replaced by a process of radical examination of alternative external choices at every stage.

'Amongst these choices from now on will be the use of futures contracts. Decisions to process additional crude oil can be protected by the sale of futures contracts until such time as physical sales are made. In the longer term, as the new contracts come into being, refinery margins will be protected by the purchase of crude oil futures and the sales of futures contracts for products.

'All of this will take time but the process of acceptance will accelerate as different sectors of the industry begin to appreciate that the integrated use of futures in business decision making should give greater freedom for business action.'

*Adviser to Czarnikow Schroder on Petroleum Futures.
†To be published in full in *Energy Economics in Britain*.

Petroleum Futures Markets as a Pricing Mechanism

*Thomas F. McKiernan**

This paper reviews the origin of oil futures markets, their current position and prospects. A specimen oil futures contract is attached as an annex. It concludes:

'Most of our business decisions concerning energy will be relative to price, supply and consumption. All of these have risk attached. Any progress towards that perfect competition model through an effective spot and futures market in the oil industry will benefit that end to the highest degree, for a competitive market promotes the public good.'

*Vice President and Director, Energy Futures Group, Drexel Burnham Lambert, New York.

PART XI

The US Potential for Gas Imports: A Supply/Demand Analysis

*Dr John T. Fraser**

US natural gas policies over the last decade have been dominated by numerous regulations and regulatory bureaucracy which have decreased the gas industry's flexibility, delayed projects, and restrained price increases. The net effect has been to discourage domestic exploration and conservation efforts—inevitably leading to a trend decline in US natural gas reserves and production.

US gas production currently meets about 95% of total US gas demand—hence, the most critical factor in assessing the need for gas imports is the *future* trend in domestic production.

The US gas market is of such magnitude that even a small gap between domestic supply and demand could provide a substantial incentive for gas imports. If, however, the US eventually attains total self-sufficiency via increased domestic supply incentives this would forestall illusory expectations about future gas imports.

This paper discusses the outlook for US gas imports against the background of US natural gas supply/demand.

Conclusions

Both LNG and Alaskan pipeline gas represent gas sources which are expensive. If LNG is priced at crude oil parity and given the financial burden of the Alaskan natural gas pipeline neither are projected to be competitive with pipeline gas imports priced to yield a burner tip price equivalent to high sulfur residual fuel oil.

Alaskan gas is expected to be a more expensive source than LNG. In 1990, when Alaskan pipeline gas is now expected to start flowing (if it does), its price is expected to be at least $11/Mcf ($ 1981). This compares unfavorably to the average city gate price of about $6.90/MMBtu projected to prevail in that year. In addition, as more of the NGPA controlled gas become deregulated in the late 1980s and early 1990s, the ability of importing pipelines to roll-in expensive Alaskan gas will decrease significantly making it less competitive with other lower cost import sources. For this reason, the Alaskan gas pipeline volumes have not been included in our base case gas supply.

Similar logic can be used to explain the backout of LNG by the cheaper Canadian and Mexican pipeline imports. Therefore, if LNG is to become a viable US gas source it must be priced competitively with pipeline imports.

In short, adequate potential exists from Canada and Mexico to back out existing and future LNG gas volumes in addition to the Alaskan gas pipeline volumes.

*Tennessee Gas Transmission. The author gratefully acknowledges the assistance of Dr Glen E. Schuler, Jr, in this endeavor. The full text of this paper including detailed methodology, the various definitions and scenarios with 25 viewgraphs, is available in the BIEE Archive.

The Influences of Higher Energy Prices on Canadian Agricultural Productivity

*M. W. Luke Chan and Dean C. Mountain**

Within the last three decades labour productivity changes in Canadian agriculture have varied considerably. However, it is only since 1973 that significant increases in energy prices have been observed in the agriculture sectors. For Canadian agriculture, the extent to which energy and other variable input and output price movements have influenced average labour productivity has never been addressed. In addition to explaining average labour productivity changes on a regional basis in terms of capital deepening changes and technical progress, this paper also endeavours to explain these productivity movements in terms of changes in output prices, material prices and energy prices. Instead of employing the traditional value added production formulation for decomposing factors influencing productivity change, this paper begins with a profit specification for the agriculture sector as a method to isolate the influence of variable input and output prices on productivity changes. The advantage of using this approach in examining the Canadian agriculture sector is that the traditional measure of the rate of technical progress fails to take account of the influence of changing energy, material and output prices.

The calculations performed in this study revealed that higher energy prices in the 1973-79 period were responsible for subtracting from labour productivity growth rates between 0.62 percentage points in British Columbia and 1.33 percentage points in Manitoba. On average, across Canada higher energy prices can account for the subtraction of approximately 0.95 percentage points from average labour productivity.

*Professor Chan is at McMaster University and Mr Mountain with Ontario Hydro, Toronto.

Two Wrongs don't make a Right: Energy R&D Policy under Carter and Reagan

*M. Krebs**

Since the 1973 oil embargo, the United States has spent billions of dollars to foster the development and utilization of new energy supply and conservation technologies. The Reagan Administration has reviewed these activities in the light of its 'free market' philosophy and is attempting to reduce the federal role in energy R&D selectively. The question at the crux of these decisions is what is the role of the Federal Government in 'commercialization' of new energy technologies for private sector consumption? The Committee on Science and Technology has been involved in providing congressional guidance for the federal energy R&D programs and has established a lengthy public record dealing with the pitfalls and promises of these programs in the last decade.

This paper represents the analysis of the Committee staff and examines the rise and fall of the commercialization concept as an instrument of policy.

The limitations of both the Carter and Reagan administrations' attitude toward commercialization are discussed using examples from the fossil energy, conservation and renewables R & D programs.

Finally, recommendations are made for a pragmatic approach to government supported research, development and demonstration programs that recognizes differing but interdependent requirements of technology choice, markets and manufacturer risk.

*Staff Director, Sub Committee on Energy Development and Application, Committee on Science and Technology, US House of Representatives.

PART XII

Natural Resource Scarcity: The Case of Australian Coal

*D. R. Gallagher and G. D. McColl**

This paper reports the results of a study relating to Australian coal resources. Its principal aim is to analyse price movements in the Australian black coal industry from the beginning of the 20th century until recently. Such an explanation may be useful as a basis for formulating inferences about the future since one purpose of such inquiries may be to develop forecasts of resource costs and prices and to comment on likely impacts on standards of living.

After examining issues raised in recent discussions about natural resource scarcity and scarcity measures, attention is focussed on available empirical evidence relating to the production and sale of Australian coal in both domestic and export markets. The final section discusses the apparent causes of the price changes which have occurred over the period examined.

From the Conclusions

The higher average export prices for Australian coal in recent years stem mainly from OPEC's intervention in the international oil market rather than a scarcity of coal resources in the world as a whole. Previous increases in such prices, in the early 1920s and late 1940s, seem to be attributable to domestic factors other than an emerging shortage of the resource itself. If increasing scarcity has had any effect at all on Australian coal prices, it has apparently been overwhelmed by other factors. The only important exception to this conclusion may be that Australia's ability to supply steaming coal demands in Japan and elsewhere has conferred some degree of monopoly pricing power as reflected in export prices, which in turn may stem from a belief that there is a growing scarcity of coal in the world. Such a view may be misguided in the light of existing and growing information about resource stocks. Further large scale investment in coal production is occurring in Australia and elsewhere; when it bears fruit in terms of additional output, it may turn out that the price rises of the 1970s were transient.

On the other hand, it may turn out that higher Australian costs will be once more instrumental in lifting the average level of real prices for Australian coal. This time the expectation is that higher coal demands following the oil price rises will not only cover higher Australian costs but also make it possible to expand exports on a large scale. But difficulties currently being experienced by Australian coal exporters in negotiating new contracts and raising prices under existing contracts suggest that this may not be easy, at least in the more immediate future.

*Centre for Applied Economic Research, the University of New South Wales.
The authors acknowledge the valuable research assistance provided by Alison Harvie, Genevieve Marshall and Thomas Mozina.

PART XIV

Share-Farming: A Possible Solution to Declining Producer Profit

*R. Deam and M. Laughton**

The structure of the crude oil market changed towards the end of 1981, characterised by

1. a two-tier price structure coming into existence with the Saudi price being lower than the rest of the OPEC producers, and
2. substantially decreased demand.

Whereas with a single-tier crude oil price structure the rise and fall in demand fell mainly on Saudi Arabia, it now falls on other producers, e.g. Kuwait, Nigeria, etc. Liftings are but one-third of their previous levels with corresponding drops in revenue. Saudi Arabia has kept its revenue reasonably constant and so now it is other producers that take the swings.

Under these circumstances, producers are tempted to reduce price (discount) to restore volume and revenue, making producers who do not do likewise suffer even greater volume and revenue loss. Such actions naturally produce friction within OPEC.

In the meantime, the refiners also have suffered difficulties with refinery profits disappearing leading, in some cases, to refiners without upgrading plant, to shut down.

This paper seeks an understanding, in a somewhat simplified manner yet, hopefully, without sacrifice of validity, of the underlying principles governing this situation leading to suggestions for possible solutions.

Refinery Operation and the Two-tier Crude Oil Price Structure

Simplified refinery economics. In order to demonstrate the essential economic principles governing refinery operation, a simple refinery model processing idealised crude oil will be considered using rounded figures for quantities and prices.

In summary, in such a situation

(a) independent refineries with low crude runs are uncertain whether to buy more or less second-tier crude, or indeed to shut down, and
(b) crude producers prevented by agreement from discounting crude prices see their liftings and revenues drop.

Both parties, crude producer and refiner, are in trouble. A hedging mechanism is required.

A Possible Solution: 'Share-Farming'

Consider the following scenario.

A producer, say Abu Dhabi, gives an independent refiner 100,000 bbls/day.

The refiner ships, refines, and markets the products to maximise his net revenue. Net revenue is defined as 'sales revenue less his transport cost, refining cost and marketing cost and taxes'. These costs do not allow any return on capital for the refiner.

A calculation is now done to determine what profit the refiner would have if crude had been supplied at 'market' price. Such a calculation might be as outlined below.

It would then seem reasonable for the producer, since he has no equity interest in the refiner to agree that the net revenue transferred to the producer should be reduced as follows.

The producer now has in times such as existed in late 1981

1. a guaranteed revenue, and
2. without investment, a share in refinery surplus profits.

Since the producer has not put in downstream investment he has no fear of a consuming government appropriating his assets. The independent refiner has a guaranteed crude supply, and a minimum profit, his 'price' being the producer getting a share of 'surplus' refiner profits.

*Queen Mary College, University of London.

Associated Gases and their Application in the Middle East Oil Producing States

*T. R. Dealtry**

The availability of associated gases for industrial and utility purposes (electricity and desalinated water) is directly linked to the crude oil production levels. Currently, well-head gases, in heat equivalent terms, of 1 million barrels of oil per day are 'flared' into the atmosphere in the Gulf Region.

The search for applications for these gases has been very intensive and the related investment decisions will have wide ranging effects internationally in the 1980s.

The paper examines the plans, comparative economies and methods employed to utilise these flared gases.

Conclusions

The second major wave of international energy impacts in less than a decade is gathering momentum. Its main features are:

1. the application of associated gases to energy and capital intensive industrial projects in the Middle East States; these experiences will lead

to the development of the many untapped natural gas fields in the Gulf Region;
2. the increasing export of gas products from Middle East States; and
3. the substantial expansion of refinery capacity in the Gulf States which has already started to change the international trading structure for refined products.

The combination of these three features of Middle East development will have an increasing effect on international trade and hence a countervailing pressure on economic development strategies in other countries. The evidence of these changes is already apparent in the world primary metals markets, refined products trading and the growing diversification in the consumption of energy products.

*R.B.A. Management Services, London.

Optimal Oil Producer Behavior considering Macrofeedbacks

*Harry D. Saunders**

Conclusions

The additional considerations of macrofeedback which are introduced into the oil producer optimization problem can be dealt with within a portfolio management framework, given a suitable macroeconomic model. The solution of the portfolio management problem results in an optimizing condition different from the usual Hotelling condition. This condition includes the Hotelling condition as a special case when macrofeedbacks to capital returns are assumed not to exist.

Strong short-run macrofeedbacks through energy demand and returns to capital appear to limit the gains to producers from adopting extreme pricing strategies. Nor do producers appear to gain much by creating uncertainty for its own sake concerning future oil prices.

Both oil producers and consuming countries appear to gain substantially if the 'long-range pricing formula' is chosen over more radical pricing strategies.

*Stanford University.

PART XV

The Current International Deadlock over Stimulating Oil Exploration in LDCs

*James L. Plummer**

Conclusions

1. There are many risks and imperfections in private markets for oil explora-

tion capital. The presence of multilateral lending institutions can make these markets more efficient by:

(a) reducing some of the political risks faced by oil companies;
(b) reducing the capital shortages facing the host government.

2. Oil proliferation financing could stimulate substantial quantities of non-OPEC supply by the mid 1990s.
3. The incentive to oil-importing countries to provide capital for oil proliferation is measured in terms of the 'indirect rate of return'—the payoff in lower future oil prices in relation to present capital outlay.
4. Even using conservative assumptions regarding costs of exploration and finding rates, the real indirect rate of return is much higher for oil proliferation than on private investments in OECD economies.
5. The present World Bank proposal for an 'energy affiliate' would not do much for oil proliferation because:

(a) only 3-5% of the lending would be for exploratory drilling;
(b) it would really be just an 'energy window' within the Bank, rather than a new institution specializing in energy;
(c) it would offer terms very similar to standard Bank terms rather than making terms (or equity participation) conditional on the outcome of the exploration.

*President, Q.E.D. Research, Inc., Palo Alto.

Energy Pricing and Institutional Arrangements: A Comparative Study of Tunisia, the Sudan and Senegal

*Peter M. Meier**

It is the thesis of this paper that within the *practical domain* of energy pricing in developing countries (as opposed to academic discussion of the principles and desirability of marginal cost pricing and the like), economic analysis must go hand in hand with institutional reality. A comprehensive pricing policy presupposes a comprehensive energy planning process, which in turn requires appropriate and effective institutional arrangements, still to be installed in many countries. Ubiquitous also is the legacy of colonial days: many countries are still wrestling with compensation arrangements to the former private owners of electric companies, issues which clearly intrude on the energy economy. Vested interests of established bureaucratic institutions, and general institutional inertia, further impede even incremental change. From the development of the argument of the importance of institutional arrangements emerges also an analytical framework for the quantitative analysis of tarification issues.

To give our discussion some real context, three African countries are examined, all almost exclusively dependent on oil, that are at distinctly different levels of macroeconomic development, and who face a somewhat different future with respect to domestic oil supply. Tunisia, currently a net oil exporter (and likely to stay so for at least another five years); the Sudan, where oil has recently been discovered, but whose development is likely to be so costly that the macroeconomic benefits remain elusive; and Senegal, where serious attempts at oil exploration are only just beginning. And despite some significant disparities, all three share that most urgent problem of deforestation and desertification.

*State University of New York at Stony Brook, Averell Harriman School of Urban & Policy Sciences and the Institute for Energy Research; and National Center for Analysis of Energy Systems, Department of Energy and Environment, Brookhaven National Laboratory.

Balance of Payments Constraints, Capital Formation, and Manufacturing Progress in Net-Oil-Importing Developing Countries

*Corazon M. Siddayao**

Focusing on the social costs of increasing oil prices, the author has argued (1) that the effective demand for oil imports of net-oil-importing developing countries (NOILDCs) is constrained by their supply of foreign exchange, (2) that the demand for oil and for capital goods for development purposes in these countries competes for this supply of foreign exchange, and (3) that one approach to relaxing the foreign exchange constraint is to increase foreign borrowing.

This paper focuses on the effect of foreign exchange constraints on the development process by concentrating on changes in the capital formation levels and manufacturing output in the NOILDCs. Special detailed attention is given to eight selected Asian countries and, where possible, data for the period 1960-1980 are used. The study finds that in these countries the incremental capital-output ratio (ICOR), which occupies an important place in the development literature and which is used as an indicator of capital productivity in individual countries, generally averaged better than the range reported for developing countries in general in the 1950s or remained within the range reported for the region for the 1950s. Using lags, more year-to-year deteriorations than improvements were noted for individual countries in the period 1973-1980.

Focusing further on the manufacturing sector and the external accounts, the study finds; (1) Manufacturing output levels provide the major explanation for levels of capital formation in both low-income South Asian and middle-income East-South Asian NOILDCs. (2) Capital formation is affected negatively by the current account balance, but this effect is stronger in the East-Southeast Asian group. (3) Oil imports are positively correlated to capital formation, more strongly in the East-Southeast Asian group than in South Asia. (4) The positive signs of the oil import coefficients fail to support fears that higher prices negatively affect investment levels; however, this result may be affected by the external debt variable. (5) Furthermore, in a three-variable estimate with external debt as the dependent variable, correlation coefficients for oil and capital imports of 0.77 and 0.35, respectively, were obtained for the middle-income NOILDCs, and 0.34 and 0.79, respectively, for the low-income countries.

*Resource Systems Institute, The East-West Center, Hawaii, USA.

Problems of Implementing Alternative Energies in the Developing World

*Markus Fritz**

The author has sent out more than 500 questionnaires in his capacity as an Energy Consultant to UNESCO to institutions in the developing world that are responsible for energy projects, in order to learn case by case about the problems encountered. The results of this survey, based on more than 200 responses, form the core of this paper.

*Max Planck Institute for Physics, Munich.

Adaptation Under Uncertainty

*C. Hope**

A simple model for the projection of national energy supply and demand is described. The model can be used to find the total costs of the national energy system, and to investigate the effects of uncertainty in key parameters. These two uses are illustrated by an application to an oil-importing developing country, and the implications for future economic growth are discussed.

*Energy Research Group, Cavendish Laboratory, Cambridge UK.

PART XVI

The Interrelationship between Energy Prices, Productivity, Industrial Location, and Regional Economic Growth

Frank Hopkins and Peter Back†*

The analysis presented in this paper summarizes research from other sources, and adds additional data to the debate on the importance of energy efficiency to the location of economic activity and regional economic growth. The initial paper on the impact potential of energy shortages and price increases, by Hogan and Manne, served to remove much of the rhetoric of the earlier debates by showing that this impact is an empirical question of the level of substitution between energy and other inputs. Their initial conclusion was that a high enough degree of substitution exixted so that relatively large increases in world energy prices could occur before economic growth rates would be significantly affected.

In a more recent paper, Jorgenson has proposed that higher energy prices result in a reduction in the growth of productivity for a large proportion of industries. The fourfold increase in world energy prices in 1973/74 and the rapid rise of prices in 1979, that has increased energy prices relative to other inputs, has been one of the major factors in reducing the overall productivity growth of the US economy.

This paper concludes that the Hogan-Manne and Jorgenson findings are consistent when analyzed in a dynamic world of uncertainty, rather than the static certain environment of the earlier authors. The unstable economic situation with its attendant high interest rates and uncertainty has constrained the economy from investing in available energy efficient technology as anticipated by Hogan and Manne. The analysis of Jorgenson follows in the absence of anticipated energy efficient investments.

*ORI, Inc.
†US Department of Energy.

The Growth Imperative Revisited

*Christopher Johnson**

'The growth imperative' was the title given to my concluding comments at the Cambridge conference of the IAEE-BIEE in June 1980. Comments then regarded as iconoclastic now look almost like a new orthodoxy. The two themes I picked out were the interdependence between oil and economic growth, and the influence which non-OPEC governments could exert on the oil market—and thus on the world economy—if they chose. These are again the two themes of my concluding comments to the 1982 Cambridge conference of the IAEE-BIEE.

The oil market is not a free market, even though some of the restrictions have been removed by the US and UK governments because of their belief in market mechanisms and the reduction of State ownership. OPEC has intervened massively and effectively in the interests of its members. There is no *prima facie* reason why countervailing intervention by other governments, in their capacity as producers or consumers, acting with the oil companies, should not be effective in reducing the oil price, or at least keeping it stable, in the interests of lower inflation and higher economic growth.

It may be noted that the oil has in fact been more stable in terms of the Special Drawing Right, over the eighteen months since the end of 1980, than in dollar terms. Dollar oil prices fell by about 12 per cent over the period, but the dollar rose by 17 per cent in terms of the SDR. SDR oil prices thus rose by 3 per cent. Pricing oil in SDRs was an idea which occurred to OPEC when the dollar was falling, but they lost interest when the dollar began to rise. The OECD countries should now seek to price oil in SDRs, thus helping to remove the destabilizing influence of the dollar exchange rate on the value of oil in other currencies.

The main suggested policy actions on the part of non-OPEC governments fall into five groups:

1. They should not encourage expectations of rising real oil prices. Official models may generate forecasts on which these expectations are based, but they are no more reliable than any other model. Official models of inflation are sometimes given a downwards bias in order to influence pay rises in a favourable direction by damping down expectations of consumer price increases. It is inconsistent of the same governments to encourage expectations of rising oil prices, which are bound to undermine expectations of a diminishing inflation rate, and convince OPEC of the wisdom of keeping oil in the ground.

2. They should seek to stabilize world oil price trends by their own price-setting activities. In June 1980, I argued that the British Government should operate the North Sea marker price in such a way as to lead world prices down rather than follow them up. Some attempt was made to do this in early 1982, although part of the original North Sea price cut was later reversed. OPEC now accounts for no more than half free world oil production, so the stereotype of OPEC as price giver and the other producers as price takers is no longer valid.
3. Governments should build up oil stocks and manage them counter-cyclically, like buffer stocks of any other commodity. Intervention in the spot market designed to eradicate the gap between spot and contract prices would discourage excessive stockbuilding or destocking, and thus remove a major cause of oil price instability. A modest margin of spare production capacity in major oilfields could be used to supplement stockpiles, perhaps at a lower cost. If Saudi Arabia has been able to vary its production in this way, major non-OPEC producers should be able to do the same.
4. Conservation of oil and switching to other fuels should be encouraged, not as a reaction to the inevitability of higher oil prices, but as a way of lowering oil prices below what they would otherwise have been. Conservation has a social return higher than the return to the investor, because, by lowering oil prices, it brings external benefits in terms of lower inflation and higher economic growth. There is thus an argument for subsidising it.
5. Governments should help to accelerate the positive response to higher oil prices, by opening up more areas to licensing, providing incentives rather than penalties through the tax system, and subsidising exploration and other research which cannot be privately funded—for example in developing countries.

*Group Economic Adviser, Lloyds Bank.

Final Panel Discussion

*J. E. Hartshorn**

The ruling theme that has come across to me at this conference has been uncertainty—plus a deep and growing disillusion with all forecasts, including one's own. The record of performance, for all techniques, is uniformly pathetic.

This inability to say much that is useful about the future might be even more worrying *if it were clear that a vast amount more investment in energy development were necessary during this decade.* In principle, such investment nowadays has such long lead times that it demands forecasting or projection, particularly of oil prices, over periods much longer than is needed in any other practical economic forecasting anywhere. But our need to know the future, unfortunately, hasn't improved our ability to do so. However, in this sense perhaps fortunately, within OECD, during this decade, we may simply not need much investment in energy development, beyond replacement, in one sense or another. If oil demand in OECD in which so far is almost the whole of effective demand for oil that really affects prices—is passing or has passed its peak, then *much of any capacity we invest in now may simply, when it comes into commission, add to one part or another of a world surplus.* And it may not remain practicable to contain all of the world's surplus capacity within OPEC. We may have to reconsider how, and whether, to invest in energy against a prospect of falling prices.

*Jensen Associates Inc.

A Kondratieff Trough

*The Rt Hon. Aubrey Jones**

I am not today in a position to present you with a number of scenarios, but I will depict one. You will have heard of the Soviet economist Kondratieff. Delving into statistics going back to the early part of the last century, he described capitalism as moving in waves—twenty-five years boom, twenty-five years depression. For this act of deviationism from the true Marxist line, which represented capitalism as heading in one straight course for an inevitable collapse, he was banished to Siberia and what since became of him I do not know.

I would like to suggest, however, that we are in the middle of a Kondratieff trough. I would place the beginning of the trough in 1968, when the students of the Sorbonne rebelled and the French Government decreed a wage increase of 15 per cent. This had repercussions throughout Europe. Then, in 1971, the US Administration cut the tie between the dollar and gold. As a result other countries became careless about inflation and the consequences for their balance of payments. There also began a bout of speculation on the foreign exchanges and in commodity markets. In 1972 the British Government, rather than incur the disgrace of a devaluation, allowed the pound to float; it inevitably floated downward. We thus created our own inflation, and the oil price increases of 1973 and 1979 were but an aggravation.

To cope with the inflation nearly all governments resorted to conventional policies which resulted in a recession. In addition there were to my mind, certain long-term trends at work making for a recession. In nearly all Western countries the apportionment of value-added between labour and capital has shifted in favour of labour, and capital has lost its dynamism.

The nature of monetary institutions has also changed. In the old days the Central Bank bought securities; the resulting cash it put out increased the reserves of the commercial banks and their ability to lend; interest rates therefore dropped. Nowadays, in the age of monetarism, when the central bank purchases securities, the money markets see the supply of money increasing and foretell inflation; they then hoist interest rates to keep pace with the inflation. I foresee therefore, a continuation of highish interest rates, a lack of investment and a persisting depression.

That is my scenario. Against it what would be my energy policy? In the short term it would be a policy of diversity: coal, oil and nuclear power. I emphasise diversity because I do not wish to be dependent on any one monopolistic source. In the longer term I would encourage research into newer forms of energy.

*President, Oxford Energy Club.

A Simple Model of the Effect of Energy Price Hikes on the World Economy

*L. G. Brookes**

Although things seem to get back to normal between the large price 'hikes', the world economy seems slow to revive. The author argues—using a simple two-dimensional model—that the explanation may be that the comparatively small increases in the real (i.e. after deducting inflation) prices of fuels may seriously understate the reduction in availability of energy that is brought about by large price hikes. He points out that energy producers do not so much put up the price of fuel as shift the supply curve. The prices at which fuels change hands are always combinations of the producers' readiness to sell and the consumers' readiness to buy. When producers put up the price they are not able to demand that the same amount shall be bought at the new price. They can only say 'if you want as much as you are getting now, you will have to pay a higher price for it. If you are not prepared to pay a higher price, then you will have to make do with less'. What happens in practice is that, after deducting inflation, the price settles down at a level somewhere between the old price and the new one demanded, with the quantity sold somewhat below what it would otherwise have been. According to the model, however, the relatively small change in real equilibrium price may conceal a substantial reduction in the availability of energy to the world economic system which does long term damage to the world economy over and above the more obvious immediate effects of price 'hikes'. Measures to mitigate the problem—although they have some beneficial effects—may actually contribute to holding up both prices and the level of demand above what they would be in the absence of such measures.

*Consultant, Bournemouth

SECTION 6

STATISTICS/TABLES

Statistics/Tables

Advance copies of the latest annual *BP Statistical Review of World Energy* were distributed to all participants at the Cambridge Energy Conference. We are very grateful to BP and to the Editor of the Review, A. F. G. Scanlan, for allowing us to reprint the abstracts in Tables A–Q. We are also grateful to Dr D. F. Lomax, Group Economic Adviser, National Westminster Bank Group, for assistance in preparing Tables R–T.

Table A. World Primary Energy Consumption 1971-1981 (MTOE)

Country/Area	*1971*	*1973*	*1975*	*1977*	*1979*	*1981*
USA	1 700.6	1 822.7	1 722.1	1 846.8	1 918.6	1.806.2
Canada	168.3	190.9	195.0	213.0	221.8	221.5
Total North America	1 868.9	2 013.6	1 917.1	2 059.8	2 140.4	2.027.7
Latin America	208.8	244.6	263.7	291.5	325.7	348.5
Total Western Hemisphere	2 077.7	2 258.2	2 180.8	2 351.3	2 466.1	2 376.2
Western Europe						
Austria	20.9	23.8	23.4	24.0	26.4	24.8
Belgium & Luxembourg	48.2	50.8	46.7	50.2	51.1	48.3
Denmark	19.8	19.7	18.2	19.9	20.1	17.1
Finland	16.1	17.9	17.4	19.4	21.6	21.3
France	162.8	186.1	171.1	186.6	192.9	188.1
Greece	11.9	15.3	17.0	19.4	17.4	16.7
Iceland	1.5	2.1	2.0	1.8	1.9	1.8
Republic of Ireland	6.8	7.4	7.1	7.7	8.7	8.2
Italy	127.9	137.9	132.9	142.7	149.7	143.8
Netherlands	64.3	76.9	71.3	75.2	78.5	69.3
Norway	25.6	28.1	28.8	28.1	31.3	28.6
Portugal	7.4	9.0	8.8	10.1	11.3	11.6
Spain	49.2	57.8	61.9	69.7	75.4	74.3
Sweden	42.8	46.2	45.1	47.6	44.1	39.4
Switzerland	22.1	23.9	23.5	25.3	24.5	26.4
Turkey	14.0	18.0	19.7	24.4	23.4	25.4
United Kingdom	213.5	224.8	204.3	211.9	221.9	195.7
West Germany	238.5	264.9	242.6	259.7	287.0	259.2
Yugoslavia	24.9	28.9	32.0	34.4	38.9	40.0
Cyprus/Gibraltar/Malta	1.2	1.4	1.1	1.4	1.5	1.4
Total Western Europe	1 119.4	1 240.9	1 174.9	1 259.5	1 327.6	1.241.4
Middle East	73.9	87.1	94.6	108.7	107.4	121.0
Africa	97.5	98.2	106.2	124.7	153.0	175.1
South Asia	94.9	103.3	105.7	118.9	130.3	154.7
South East Asia	106.1	127.4	136.8	155.3	190.3	200.3
Japan	301.4	347.7	330.5	348.1	369.9	353.6
Australasia	61.9	66.6	71.5	80.3	86.1	89.0
USSR	788.9	874.1	969.8	1 060.6	1 134.9	1 198.1
Eastern Europe	344.6	357.2	387.7	423.6	441.8	439.8
China	320.6	359.5	406.2	455.3	532.5	499.6
Total Eastern Hemisphere	3 309.2	3 662.0	3 783.9	4 135.0	4 472.9	4 472.6
World (excl. USSR E. Europe & China)	3 932.8	4 329.4	4 201.0	4 546.8	4 830.7	4 711.3
World	5 386.9	5 920.2	5 964.7	6 486.3	6 939.0	6 848.8

Table B. World Primary Energy Consumption 1981 by Sectors (MTOE)

Country/Area	*Oil*	*Natural Gas*	*Coal*	*Water Power*	*Nuclear Energy*	*Total*
USA	743.2	509.4	406.3	74.0	73.3	1 806.2
Canada	81.6	48.1	22.9	59.2	9.7	221.5
Total North America	**824.8**	**557.5**	**429.2**	**133.2**	**83.0**	**2 027.7**
Latin America	227.8	54.4	16.9	48.6	0.8	348.5
Total Western Hemisphere	**1 052.6**	**611.9**	**446.1**	**181.8**	**83.8**	**2 376.2**
Western Europe						
Austria	10.9	4.0	3.1	6.8	–	24.8
Belgium & Luxembourg	25.6	9.5	10.3	0.1	2.8	48.3
Denmark	11.7	–	5.4	–	–	17.1
Finland	12.0	0.6	1.8	3.4	3.5	21.3
France	99.3	24.7	26.2	15.8	22.1	188.1
Greece	11.9	–	3.9	0.9	–	16.7
Iceland	0.5	–	–	1.3	–	1.8
Republic of Ireland	5.4	0.9	1.7	0.2	–	8.2
Italy	95.5	22.8	13.5	10.7	1.3	143.8
Netherlands	35.2	28.8	4.3	–	1.0	69.3
Norway	7.5	–	0.5	20.6	–	28.6
Portugal	8.9	–	0.4	2.3	–	11.6
Spain	48.0	1.9	16.4	6.2	1.8	74.3
Sweden	21.8	–	1.4	9.9	6.3	39.4
Switzerland	11.9	0.8	0.7	9.3	3.7	26.4
Turkey	15.4	–	7.8	2.2	–	25.4
United Kingdom	74.6	42.0	69.7	1.3	8.1	195.7
West Germany	117.6	41.2	82.7	5.8	11.9	259.2
Yugoslavia	14.4	3.7	14.6	7.3	–	40.0
Cyprus/Gibraltar/Malta	1.4	–	*a*	–	–	1.4
Total Western Europe	**629.5**	**180.9**	**264.4**	**104.1**	**62.5**	**1 241.4**
Middle East	84.7	35.3	*a*	1.0	–	121.0
Africa	75.9	18.4	67.1	13.7	–	175.1
South Asia	43.4	8.9	88.7	12.8	0.9	154.7
South East Asia	123.7	7.3	57.2	8.6	3.5	200.3
Japan	224.3	24.2	63.2	20.7	21.2	353.6
Australasia	36.3	11.6	31.5	9.6	–	89.0
USSR	444.1	353.7	336.8	48.5	15.0	1 198.1
Eastern Europe	102.4	69.4	258.0	5.8	4.2	439.8
China	84.8	10.4	394.2	10.2	–	499.6
Total Eastern Hemisphere	**1 849.1**	**720.1**	**1 561.1**	**235.0**	**107.3**	**4 472.6**
World (excl. USSR E. Europe & China)	**2 270.4**	**898.5**	**1 018.2**	**352.3**	**171.9**	**4 711.3**
World	**2 901.7**	**1 332.0**	**2 007.2**	**416.8**	**191.1**	**6 848.8**

[a] Less than 0.05 million tonnes oil equivalent.

Table C. World Oil Production 1971-1981 (million tonnes)

Country/Area	1971	1973	1975	1977	1979	1981
USA – Crude Oil	469.9	457.3	415.9	409.4	424.7	425.2
– Natural Gas Liquids	60.1	61.7	58.0	57.4	56.2	56.4
Sub-total	530.0	519.0	473.9	466.8	480.9	481.6
Canada	76.6	102.3	83.5	75.7	83.8	73.7
Total North America	**606.6**	**621.3**	**557.4**	**542.5**	**564.7**	**555.3**
Latin America						
Argentina	22.1	22.0	20.3	22.5	24.5	25.8
Brazil	8.3	8.1	8.4	8.0	8.3	10.6
Colombia	11.0	9.4	8.1	7.0	7.5	6.8
Ecuador	0.2	10.2	7.9	9.1	10.5	10.7
Mexico	23.8	26.9	39.3	53.7	80.8	128.3
Trinidad	6.7	8.6	11.2	11.5	10.8	12.0
Venezuela	187.7	179.0	125.3	119.5	125.4	112.1
Other Latin America	6.4	7.9	7.1	7.3	12.5	12.6
Total Latin America	**266.2**	**272.1**	**227.6**	**238.6**	**280.3**	**318.9**
Total Western Hemisphere	**872.8**	**893.4**	**785.0**	**781.1**	**845.0**	**874.2**
Western Europe						
Austria	2.5	2.6	2.0	1.8	1.8	1.3
France	1.9	1.3	1.0	1.0	1.2	1.7
Italy	1.4	1.0	1.0	1.1	1.8	1.7
Norway	0.3	1.8	9.3	13.5	18.8	24.9
Turkey	3.5	3.5	3.1	2.7	2.8	2.4
United Kingdom	0.1	0.1	1.4	37.5	77.9	89.4
West Germany	7.4	6.6	5.7	5.4	4.8	4.5
Yugoslavia	3.0	3.3	3.9	4.0	4.1	4.3
Other Western Europe	1.8	2.4	3.4	2.7	3.3	3.8
Total Western Europe	**21.9**	**22.6**	**30.8**	**69.7**	**116.5**	**134.0**
Middle East						
Abu Dhabi	44.9	62.6	67.3	80.0	70.2	54.8
Dubai	6.2	10.8	12.6	15.8	17.6	17.8
Iran	227.0	293.2	267.7	283.5	158.1	65.5
Iraq	83.5	99.0	111.0	115.2	170.6	44.2
Kuwait	143.8	140.4	94.0	91.5	113.2	48.1

Table C. *(continued)*

Country/Area	*1971*	*1973*	*1975*	*1977*	*1979*	*1981*
Neutral Zone	28.3	27.6	25.8	20.8	29.4	19.2
Oman	14.4	14.7	17.1	17.1	14.8	15.8
Qatar	20.5	27.3	21.0	21.6	24.7	19.5
Saudi Arabia	225.0	367.9	343.9	455.0	469.9	491.6
Sharjah	—	—	1.9	1.4	0.7	0.5
Other Middle East	9.1	9.0	12.8	12.2	11.7	11.0
Total Middle East	**807.7**	**1 052.5**	**975.1**	**1 114.1**	**1 080.9**	**788.0**
Africa						
Algeria	36.5	51.2	47.5	53.5	56.9	46.4
Egypt	21.0	13.0	14.8	21.0	26.5	32.1
Libya	133.1	104.9	71.3	99.4	100.7	53.9
Other North Africa	4.2	3.8	4.5	4.6	5.4	5.5
Gabon	5.8	7.5	11.2	11.1	10.2	7.6
Nigeria	74.7	100.1	88.8	103.6	114.2	71.3
Other West Africa	5.8	9.5	10.4	11.5	13.1	16.5
Total Africa	**281.1**	**290.0**	**248.5**	**304.7**	**327.0**	**233.3**
South Asia	8.6	9.0	9.6	11.9	14.8	13.5
South East Asia						
Brunei	7.5	11.6	9.4	10.9	11.9	8.2
Indonesia	44.1	66.0	64.6	83.5	78.8	79.3
Other South East Asia	3.5	4.4	4.5	9.4	15.8	12.4
Total South East Asia	**55.1**	**82.0**	**78.5**	**103.8**	**106.5**	**99.9**
Japan	0.7	0.7	0.6	0.6	0.5	0.4
Australasia	15.0	18.5	19.9	21.4	21.9	19.8
USSR	377.1	429.0	490.8	545.8	586.0	609.0
Eastern Europe	18.2	19.2	20.0	20.0	20.0	17.2
China	36.7	54.8	74.3	93.6	106.1	101.1
Total Eastern Hemisphere	**1 622.1**	**1 978.3**	**1 948.1**	**2 285.6**	**2 380.2**	**2 016.1**
World (excl. USSR E. Europe & China)	**2 062.9**	**2 368.7**	**2 148.0**	**2 407.3**	**2 513.1**	**2 163.1**
World	**2 494.9**	**2 871.7**	**2 733.1**	**3 066.7**	**3 225.2**	**2 890.3**

Table D. World Oil Consumption 1971-1981 (million tonnes)

Country/Area	1971	1973	1975	1977	1979	1981
USA	719.3	818.0	765.9	865.9	868.0	743.2
Canada	75.8	83.7	83.1	85.6	90.1	81.6
Total North America	**795.1**	**901.7**	**849.0**	**951.5**	**958.1**	**824.8**
Latin America	143.5	168.3	176.0	193.4	214.2	227.8
Total Western Hemisphere	**938.6**	**1 070.0**	**1 025.0**	**1 144.9**	**1 172.3**	**1 052.6**
Western Europe						
Austria	10.1	11.8	10.7	11.1	12.5	10.9
Belgium & Luxembourg	28.4	31.5	26.5	28.0	27.6	25.6
Denmark	18.4	17.9	15.7	16.6	15.9	11.7
Finland	11.1	13.3	11.9	12.5	13.3	12.0
France	102.8	127.3	110.4	114.6	118.3	99.3
Greece	7.4	10.0	9.9	10.8	12.4	11.9
Iceland	0.5	0.7	0.6	0.6	0.6	0.5
Republic of Ireland	4.5	5.4	5.2	5.7	6.3	5.4
Italy	93.8	103.6	94.5	96.1	103.2	95.5
Netherlands	36.0	41.3	34.8	37.6	41.3	35.2
Norway	8.2	8.6	8.0	8.9	9.0	7.5
Portugal	5.4	6.3	6.8	7.1	8.1	8.9
Spain	30.9	39.1	42.7	45.5	49.1	48.0
Sweden	28.2	29.4	26.6	28.1	28.4	21.8
Switzerland	13.3	14.7	12.5	13.1	12.9	11.9
Turkey	9.0	12.4	13.4	16.6	14.7	15.4
United Kingdom	104.3	113.2	92.0	92.0	94.5	74.6
West Germany	133.5	149.7	128.9	137.1	147.0	117.6
Yugoslavia	9.0	11.3	12.2	13.9	15.8	14.4
Cyprus/Gibraltar/Malta	1.2	1.4	1.1	1.4	1.5	1.4
Total Western Europe	**656.0**	**748.9**	**664.4**	**697.3**	**732.4**	**629.5**
Middle East	54.0	62.2	66.8	78.9	75.4	84.7
Africa	44.4	49.5	51.5	59.3	65.9	75.9
South Asia	27.9	31.3	30.1	34.5	37.7	43.4
South East Asia	63.9	77.6	81.2	95.8	119.2	123.7
Japan	219.7	269.1	244.0	260.4	265.1	224.3
Australasia	31.3	34.8	35.1	38.0	38.1	36.3
USSR	279.2	325.7	375.1	399.6	427.0	444.1
Eastern Europe	61.3	75.1	83.3	96.2	101.1	102.4
China	36.7	53.8	68.3	82.0	91.1	84.8
Total Eastern Hemisphere	**1 474.4**	**1 728.0**	**1 699.8**	**1 842.0**	**1 953.0**	**1 849.1**
World (excl. USSR E. Europe & China)	**2 035.8**	**2 343.4**	**2 198.1**	**2 409.1**	**2 506.1**	**2 270.4**
World	**2 413.0**	**2 798.0**	**2 724.8**	**2 986.9**	**3 125.3**	**2 901.7**

Table E. Main Oil Product Consumption (including Bunkers) 1971-1981 (million tonnes)

Country/Area	1971	1973	1975	1977	1979	1981
USA						
Gasolines	283.5	312.7	304.5	329.5	323.5	299.3
Middle Distillates	177.5	200.7	190.0	224.2	231.2	199.2
Fuel Oil	120.3	148.3	129.4	160.8	148.5	109.0
Others	138.0	156.3	142.0	151.4	164.8	135.7
Total	**719.3**	**818.0**	**765.9**	**865.9**	**868.0**	**743.2**
Canada						
Gasolines	23.1	26.7	29.0	29.8	33.8	31.6
Middle Distillates	23.7	26.2	26.1	26.8	27.1	25.9
Fuel Oil	16.7	17.4	16.1	15.8	15.0	11.5
Others	12.3	13.4	11.9	13.2	14.2	12.6
Total	**75.8**	**83.7**	**83.1**	**85.6**	**90.1**	**81.6**
Western Europe						
Gasolines	114.3	132.6	126.8	132.8	148.7	139.1
Middle Distillates	214.6	248.7	222.9	244.3	254.6	218.9
Fuel Oil	241.9	269.9	230.0	227.1	231.8	184.7
Others	85.2	97.7	84.7	93.1	97.3	86.8
Total	**656.0**	**748.9**	**664.4**	**697.3**	**732.4**	**629.5**
Japan						
Gasolines	34.8	42.8	39.4	44.6	46.2	39.8
Middle Distillates	36.4	50.9	50.4	55.7	61.5	60.3
Fuel Oil	120.2	142.6	125.1	128.9	123.7	92.4
Others	28.3	32.8	29.1	31.2	33.7	31.8
Total	**219.7**	**269.1**	**244.0**	**260.4**	**265.1**	**224.3**
Australasia						
Gasolines	10.0	11.3	12.2	12.9	13.6	13.6
Middle Distillates	8.0	8.5	9.7	10.8	10.7	10.4
Fuel Oil	8.8	10.2	8.6	9.1	8.9	7.6
Others	4.5	4.8	4.6	5.2	4.9	4.7
Total	**31.3**	**34.8**	**35.1**	**38.0**	**38.1**	**36.3**
Rest of World (excl. USSR, E.Europe & China)						
Gasolines	57.5	70.8	78.5	80.3	92.6	101.8
Middle Distillates	99.7	113.8	129.1	154.0	164.8	182.0
Fuel Oil	129.9	150.0	140.5	163.3	185.5	195.9
Others	46.6	54.3	57.5	64.3	69.5	75.8
Total	**333.7**	**388.9**	**405.6**	**461.9**	**512.4**	**555.5**
World (excl. USSR, E.Europe & China)						
Gasolines	523.2	596.9	590.4	629.9	658.4	625.2
Middle Distillates	559.9	648.8	628.2	715.8	749.9	696.7
Fuel Oil	637.8	738.4	649.7	705.0	713.4	601.1
Others	314.9	359.3	329.8	358.4	384.4	347.4
Total	**2 035.8**	**2 343.4**	**2 198.1**	**2 409.1**	**2 506.1**	**2 270.4**

Note: 'Others' represent refinery gas, LPGs, solvents, petroleum coke, lubricants, bitumen, wax and refinery fuel and loss.

Table F. World Oil Refining Capacities at End Year 1971-1981
(thousand barrels per calendar day)

Country/Area	*1971*	*1973*	*1975*	*1977*	*1979*	*1981*
USA	13 290	14 360	15 235	16 945	17 920	18 290
Canada	1 635	1 875	2 080	2 220	2 315	2 155
Total North America	**14 925**	**16 235**	**17 315**	**19 165**	**20 235**	**20 445**
Latin America						
Argentina	630	630	700	660	660	690
Brazil	565	795	1 020	1 175	1 230	1 440
Mexico	590	735	785	980	1 395	1 395
Netherlands Antilles	800	810	810	840	840	840
Trinidad	440	460	460	465	465	455
Venezuela	1 295	1 475	1 445	1 445	1 445	1 435
Other Latin America	1 600	2 040	2 300	2 605	2 765	2 760
Total Latin America	**5 920**	**6 945**	**7 520**	**8 170**	**8 800**	**9 015**
Total Western Hemisphere	**20 845**	**23 180**	**24 835**	**27 335**	**29 035**	**29 460**
Western Europe						
Belgium	835	860	975	1 055	1 020	1 000
France	2 420	3 165	3 320	3 465	3 420	3 245
Italy	3 540	3 790	4 345	4 270	4 205	4 105
Netherlands	1 560	1 810	1 950	1 835	1 815	1 715
Spain	780	995	1 180	1 210	1 475	1 550
United Kingdom	2 390	2 825	2 950	2 655	2 460	2 460
West Germany	2 535	2 920	3 115	3 065	3 040	2 955
Other Western Europe	1 810	2 175	2 810	2 925	3 080	3 335
Total Western Europe	**15 870**	**18 540**	**20 645**	**20 480**	**20 515**	**20 365**

Table F. *(continued)*

Country/Area	*1971*	*1973*	*1975*	*1977*	*1979*	*1981*
Middle East						
Bahrain	250	250	250	250	250	250
Iran	600	645	715	965	1 045	640
Iraq	110	115	185	185	265	265
Kuwait	540	525	565	575	575	575
Neutral Zone	75	75	75	80	80	80
Saudi Arabia	425	425	465	630	635	585
Southern Yemen	160	160	160	145	145	145
Other Middle East	250	330	335	415	545	690
Total Middle East	**2 410**	**2 525**	**2 750**	**3 245**	**3 540**	**3 230**
Africa	780	1 070	1 270	1 505	2 020	2 160
South Asia	630	655	735	785	825	890
South East Asia						
Indonesia	415	415	415	525	535	515
Singapore	430	1 040	1 040	1 040	1 040	1 020
Other South East Asia	840	1 145	1 290	1 610	1 850	1 975
Total South East Asia	**1 685**	**2 600**	**2 745**	**3 175**	**3 425**	**3 510**
Japan	3 655	4 815	5 215	5 285	5 285	5 675
Australasia	700	740	760	760	800	820
USSR, E.Europe & China	8 720	10 405	12 250	13 320	14 650	15 990
Total Eastern Hemisphere	**34 450**	**41 350**	**46 370**	**48 555**	**51 060**	**52 640**
World (excl. USSR, E.Europe & China)	**46 575**	**54 125**	**58 955**	**62 570**	**65 445**	**66 110**
World	**55 295**	**64 530**	**71 205**	**75 890**	**80 095**	**82 100**

Conversion factor: crude oil. 1 million barrels/day = 49.8 million tonnes/year.

Table G. World Refinery Crude Oil Throughputs 1971-1981

(thousand barrels daily)

Country/Area	*1971*	*1973*	*1975*	*1977*	*1979*	*1981*
USA	11 320	12 555	12 475	14 625	14 465	12 475
Canada	1 395	1 657	1 700	1 805	1 960	1 965
Latin America	5 295	5 970	5 515	5 715	6 220	6 590
Western Europe	13 145	15 065	12 415	13 550	14 160	11 870
Middle East	2 240	2 170	2 075	2 240	2 365	2 230
Africa	765	930	1 005	1 165	1 355	1 565
South Asia	545	590	485	605	675	745
South East Asia	1 265	1 705	1 545	2 030	2 450	2 430
Japan	3 525	4 430	4 055	4 285	4 340	3 665
Australasia	565	620	635	680	700	655
USSR, E.Europe & China	7 845	9 365	11 025	11 930	13 230	13 420
World	**47 905**	**55 075**	**52 930**	**58 630**	**61 920**	**57 610**

Table H. World Crude Oil Trade 1971-1981 (thousand barrels daily)

Country/Area	*1971*	*1973*	*1975*	*1977*	*1979*	*1981*
Imports						
USA	3 930	6 255	6 025	8 710	8 410	5 950
Western Europe	13 520	15 405	12 610	13 295	13 080	10 245
Japan	4 720	5 480	4 945	5 510	5 605	4 460
Rest of World	5 970	6 890	6 755	7 240	8 255	8 000
World	**28 140**	**34 030**	**30 335**	**34 755**	**35 350**	**28 655**
Exports						
USA	220	230	210	250	485	605
Canada	835	1 340	800	540	455	455
Latin America	3 655	3 885	3 175	3 115	3 645	4 380
Middle East	15 250	19 990	18 505	20 730	20 435	14 605
North Africa	3 880	3 400	2 405	3 295	3 425	2 175
West Africa	1 625	2 010	1 975	2 370	2 645	1 730
South East Asia	930	1 370	1 455	1 915	1 915	1 660
USSR, E.Europe & China	1 270	1 370	1 470	1 975	1 740	2 195
Other Eastern Hemisphere	475	435	340	565	605	850
World	**28 140**	**34 030**	**30 335**	**34 755**	**35 350**	**28 655**

Table I. Inter-Area Total Oil Movements 1981 **(million tonnes)**

To	*USA*	*Latin America*	*Western Europe*	*Southeast Asia*	*Japan*	*Total Exports*
From						
USA	–	13.1	7.2	2.1	2.1	30.8
Canada	21.2	0.1	0.7	–	–	22.5
Latin America	96.2	10.8	44.3	–	7.6	217.7
Western Europe	26.1	–	–	0.5	0.3	37.6
Middle East	61.5	69.2	289.7	87.3	148.2	725.4
North Africa	32.5	3.5	67.0	–	3.5	107.9
West Africa	35.8	14.2	32.6	–	1.1	85.9
East & South Africa	–	–	–	0.3	0.1	0.8
South Asia	–	–	–	0.1	0.3	2.2
South East Asia	19.6	–	2.4	–	47.3	82.5
Japan	–	–	–	0.3	–	0.4
Australasia	0.1	–	–	0.5	–	0.6
USSR, E.Europe & China	0.9	9.2	69.5	11.9	9.5	108.8
Total Imports	293.9	120.1	513.4	103.0	220.0	1 423.1

Table J. World Natural Gas Production 1971-1981 (MTOE)

Country/Area	*1971*	*1973*	*1975*	*1977*	*1979*	*1981*
USA	560.6	554.4	490.7	488.9	501.6	499.2
Canada	58.1	69.6	68.9	72.7	75.1	68.0
Total North America	**618.7**	**624.0**	**559.6**	**561.6**	**576.7**	**567.2**
Latin America						
Argentina	5.9	6.1	6.9	6.8	6.5	7.2
Bolivia	0.1	1.5	1.5	1.6	2.0	2.4
Chile	3.2	3.7	3.1	3.4	3.1	3.2
Mexico	12.2	13.8	14.9	18.5	23.4	32.1
Trinidad	1.7	1.6	1.6	1.5	2.9	2.2
Venezuela	9.4	11.7	11.5	13.4	14.7	15.5
Other Latin America	2.1	2.8	3.2	3.7	4.2	3.7
Total Latin America	**34.6**	**41.2**	**42.7**	**48.9**	**56.8**	**66.3**
Total Western Hemisphere	**653.3**	**665.2**	**602.3**	**610.5**	**633.5**	**633.5**
Western Europe						
France	6.7	7.0	6.9	7.1	7.2	6.6
Italy	11.4	13.8	13.1	11.6	12.1	11.6
Netherlands	33.1	53.6	68.7	73.3	70.8	65.0
Norway	—	—	0.2	2.8	19.4	22.7
United Kingdom	16.1	25.4	31.9	35.2	34.2	31.9
West Germany	11.5	14.5	13.7	14.4	15.4	16.7
Other Western Europe	2.8	3.4	3.7	3.9	3.9	3.1
Total Western Europe	**81.6**	**117.7**	**138.2**	**148.3**	**163.0**	**157.6**
Middle East						
Abu Dhabi	0.9	1.1	1.0	2.9	5.6	5.4
Iran	14.1	17.8	19.7	18.9	18.0	5.7
Kuwait	5.6	4.7	4.7	5.4	7.8	5.2
Saudi Arabia	2.4	4.1	5.1	7.2	10.5	13.2
Other Middle East	2.8	4.7	6.2	6.2	9.7	8.5
Total Middle East	**25.8**	**32.4**	**36.7**	**40.6**	**51.6**	**38.0**

Table J. *(continued)*

Country/Area	*1971*	*1973*	*1975*	*1977*	*1979*	*1981*
Africa						
Algeria	2.7	4.3	8.6	7.8	23.4	19.6
Libya	0.6	2.8	3.1	3.5	6.1	2.9
Nigeria	0.2	0.3	0.4	0.5	1.2	0.7
Other Africa	0.2	0.6	0.6	1.0	1.6	1.8
Total Africa	**3.7**	**8.0**	**12.7**	**12.8**	**32.3**	**25.0**
South Asia						
Pakistan	2.7	4.0	4.2	5.3	5.1	6.1
Other South Asia	3.5	4.6	4.6	5.2	5.7	7.0
Total South Asia	**6.2**	**8.6**	**8.8**	**10.5**	**10.8**	**13.1**
South East Asia						
Indonesia	3.1	4.6	2.1	5.1	14.2	16.9
Other South East Asia	1.2	4.4	7.4	10.3	10.0	10.9
Total South East Asia	**4.3**	**9.0**	**9.5**	**15.4**	**24.2**	**27.8**
Japan	2.4	2.6	2.2	2.5	2.2	1.8
Australasia	2.5	4.5	5.5	8.4	9.8	12.0
USSR	191.3	212.9	260.4	311.7	353.6	411.8
Eastern Europe	34.0	40.9	46.7	46.8	46.2	46.3
China	4.4	6.6	8.9	10.8	12.5	11.6
Total Eastern Hemisphere	**356.2**	**443.2**	**529.6**	**607.8**	**706.2**	**745.0**
World (excl. USSR, E.Europe & China)	**779.8**	**848.0**	**815.9**	**849.0**	**927.4**	**908.8**
World	**1 009.5**	**1 108.4**	**1 131.9**	**1 218.3**	**1 339.7**	**1 378.5**

Table K. World Natural Gas Consumption 1971-1981 (MTOE)

Country/Area	1971	1973	1975	1977	1979	1981
USA	584.2	572.3	508.7	502.3	520.8	509.4
Canada	34.9	41.8	43.1	45.9	50.1	48.1
Total North America	**619.1**	**614.1**	**551.8**	**548.2**	**570.9**	**557.5**
Latin America	32.6	36.5	39.2	39.6	49.1	54.4
Total Western Hemisphere	**651.7**	**650.6**	**591.0**	**587.8**	**620.0**	**611.9**
Western Europe						
Austria	2.9	3.4	3.6	4.2	4.3	4.0
Belgium & Luxembourg	5.8	8.2	9.6	10.1	10.3	9.5
Denmark	—	—	—	—	—	—
Finland	—	—	0.7	0.7	0.8	0.6
France	11.1	15.7	17.0	20.4	23.3	24.7
Greece	—	—	—	—	—	—
Iceland	—	—	—	—	—	—
Republic of Ireland	—	—	—	—	0.3	0.9
Italy	12.5	14.4	18.0	21.6	22.9	22.8
Netherlands	24.6	32.2	33.2	33.4	33.1	28.8
Norway	—	—	—	—	—	—
Portugal	—	—	—	—	—	—
Spain	0.4	1.0	1.3	1.4	1.4	1.9
Sweden	—	—	—	—	—	—
Switzerland	*a*	0.2	0.5	0.6	0.8	0.8
Turkey	—	—	—	—	—	—
United Kingdom	18.1	26.1	32.9	36.9	41.9	42.0
West Germany	16.9	27.0	34.4	38.9	46.2	41.2
Yugoslavia	1.1	1.7	2.2	1.7	2.3	3.7
Cyprus/Gibraltar/Malta	—	—	—	—	—	—
Total Western Europe	**93.4**	**19.9**	**153.4**	**169.9**	**187.6**	**180.9**
Middle East	19.0	24.1	26.2	28.8	31.0	35.3
Africa	1.7	3.2	4.3	6.9	16.5	18.4
South Asia	6.0	8.0	8.1	9.4	6.4	8.9
South East Asia	2.3	3.6	4.1	4.6	6.4	7.3
Japan	3.8	5.3	7.7	10.9	20.3	24.2
Australasia	2.2	3.9	4.9	7.4	8.7	11.6
USSR	173.8	198.8	230.0	271.2	307.0	353.7
Eastern Europe	38.5	42.1	51.3	58.4	61.5	69.4
China	4.3	6.4	8.7	10.9	12.4	10.4
Total Eastern Hemisphere	**345.0**	**425.3**	**498.7**	**578.4**	**657.8**	**720.1**
World (excl. USSR E. Europe & China)	**780.1**	**828.6**	**799.7**	**825.7**	**896.9**	**898.5**
World	**996.7**	**1 075.9**	**1 089.7**	**1 166.2**	**1 277.8**	**1 332.0**

[a] Less than 0.05 million tonnes oil equivalent.

Table L. World Reserves of Oil and Gas—'Published Proved'—at End 1981

	Oil			*Natural Gas*		
Country/Area	*Thousand Million Tonnes*	*Share of Total*	*Thousand Million Barrels*	*Trillion[a] Cubic Feet*	*Share of Total*	*Trillion[a] Cubic Metres*
USA	4.7	5.4%	36.5	198.0	6.8%	5.6
Canada	1.0	1.2%	7.9	90.5	3.1%	2.6
Total North America	5.7	6.6%	44.4	288.5	9.9%	8.2
Latin America	11.9	12.5%	85.0	176.3	6.1%	5.0
Total Western Hemisphere	17.6	19.1%	129.4	464.8	16.0%	13.2
Western Europe	3.4	3.7%	25.2	152.7	5.2%	4.3
Middle East	49.3	53.5%	362.6	761.9	26.2%	21.6
Africa	7.5	8.3%	56.2	211.7	7.3%	6.0
USSR	8.6	9.3%	63.0	1 160.0	39.8%	32.8
Eastern Europe	0.4	0.4%	2.7	8.8	0.3%	0.2
China	2.7	2.9%	19.9	24.4	0.8%	0.7
Other Eastern Hemisphere	2.6	2.8%	19,2	127.6	4.4%	3.6
Total Eastern Hemisphere	74.5	80.9%	548.8	2 447.1	84.0%	69.2
World (excl. USSR, E.Europe & China)	80.4	87.4%	592.6	1 718.7	59.1%	48.7
World	92.1	100.0%	678.2	2 911.9	100.0%	82.4

[a]Trillion: 10^{12}; one million million.

Source of data: Canada — Canadian Petroleum Association.
All other areas — Estimates published by the *Oil and Gas Journal* (Worldwide Oil issue: 28th December, 1981), plus an estimate of natural gas liquids for the USA.

Notes:
1. Proved crude oil and natural gas reserves are generally taken to be the volume of oil and gas remaining in the ground which geological and engineering information indicate with reasonable certainty to be recoverable in the future from known reservoirs under existing economic and operating conditions.
2. The recovery factor, i.e. the relationship between proved reserves and total oil or total gas in place varies according to local conditions and can vary in time with economic and technological changes.
3. For the USA and Canada the oil data include natural gas liquids which it is estimated can be recovered from proved natural gas reserves.
4. The data exclude shale oil and tar sands.
5. Percentages are based on volume.

Table M. World Coal Consumption 1971-1981 (MTOE)

Country/Area	*1971*	*1973*	*1975*	*1977*	*1979*	*1981*
USA	316.3	335.0	322.9	351.9	380.7	406.3
Canada	16.1	15.6	15.5	23.4	18.2	22.9
Total North America	332.4	350.6	338.4	375.3	398.9	429.2
Latin America	10.5	11.7	14.2	14.1	16.0	16.9
Total Western Hemisphere	342.9	362.3	352.6	389.4	414.9	446.1
Western Europe						
Austria	3.5	3.6	3.2	2.9	3.1	3.1
Belgium & Luxembourg	13.7	11.0	8.9	9.4	10.6	10.3
Denmark	1.4	1.8	2.5	3.3	4.2	5.4
Finland	2.2	2.0	1.8	2.6	3.2	1.8
France	33.8	29.5	26.5	29.8	28.5	26.2
Greece	3.7	4.7	6.5	8.1	4.0	3.9
Iceland	*a*	*a*	*a*	—	—	—
Republic of Ireland	2.1	1.8	1.7	1.8	1.9	1.7
Italy	9.5	9.0	9.8	9.6	10.7	13.5
Netherlands	3.6	3.1	2.5	3.2	3.3	4.3
Norway	0.9	0.5	0.6	0.5	0.5	0.5
Portugal	0.4	0.6	0.4	0.4	0.4	0.4
Spain	8.6	8.5	9.2	10.6	10.9	16.4
Sweden	0.9	0.8	0.9	1.0	1.7	1.4
Switzerland	0.4	0.2	0.1	0.3	0.5	0.7
Turkey	4.3	4.6	4.9	5.5	6.5	7.8
United Kingdom	82.9	78.4	71.9	73.4	76.1	69.7
West Germany	83.8	82.1	70.7	71.7	79.8	82.7
Yugoslavia	10.7	11.4	12.6	12.5	13.9	14.6
Cyprus/Gibraltar/Malta	*a*	*a*	*a*	*a*	*a*	*a*
Total Western Europe	266.4	253.6	234.7	246.6	259.8	264.4
Middle East	*a*	*a*	*a*	*a*	*a*	*a*
Africa	44.4	38.6	43.1	47.3	58.1	67.1
South Asia	52.4	55.0	59.1	64.0	74.9	88.7
South East Asia	33.6	39.8	44.2	47.2	55.1	57.2
Japan	53.9	53.7	54.4	52.5	50.4	63.2
Australasia	21.3	22.9	26.1	26.8	30.1	31.5
USSR	302.3	315.0	326.2	341.2	342.5	336.8
Eastern Europe	241.2	235.0	246.9	260.7	270.0	258.0
China	273.2	292.5	321.8	354.4	420.0	394.2
Total Eastern Hemisphere	1 288.7	1 306.1	1 356.5	1 440.7	1 560.9	1 561.1
World (excl. USSR E. Europe & China)	814.9	825.9	814.2	873.8	943.3	1 018.2
World	1 631.6	1 668.4	1 709.1	1 830.1	1 975.2	2 007.2

[a] Less than 0.05 million tonnes oil equivalent.

Table N. World Water Power Consumption 1971-1981 (MTOE)

Country/Area	*1971*	*1973*	*1975*	*1977*	*1979*	*1981*
USA	70.9	75.6	80.2	58.9	79.8	74.0
Canada	40.5	46.1	50.3	51.2	54.8	59.2
Total North America	**111.4**	**121.7**	**130.5**	**110.1**	**134.6**	**133.2**
Latin America	22.2	28.1	33.6	44.0	45.6	48.6
Total Western Hemisphere	**133.6**	**149.8**	**164.1**	**154.1**	**180.2**	**181.8**
Western Europe						
Austria	4.4	5.0	5.9	5.8	6.5	6.8
Belgium & Luxembourg	0.3	0.1	0.1	0.1	0.1	0.1
Denmark	—	—	—	—	—	—
Finland	2.8	2.6	3.0	3.0	2.7	3.4
France	12.8	10.6	13.3	16.7	14.5	15.8
Greece	0.8	0.6	0.6	0.5	1.0	0.9
Iceland	1.0	1.4	1.4	1.2	1.3	1.3
Republic of Ireland	0.2	0.2	0.2	0.2	0.2	0.2
Italy	11.2	10.1	9.7	14.5	11.7	10.7
Netherlands	—	—	—	—	—	—
Norway	16.5	19.0	20.2	18.7	21.8	20.6
Portugal	1.6	2.1	1.6	2.6	2.8	2.3
Spain	8.6	7.5	6.8	10.5	12.3	6.2
Sweden	13.7	15.5	14.6	13.6	10.4	9.9
Switzerland	7.7	7.2	8.5	9.2	7.9	9.3
Turkey	0.7	1.0	1.4	2.3	2.2	2.2
United Kingdom	1.1	1.2	1.2	1.2	1.3	1.3
West Germany	2.9	3.3	3.6	3.7	4.2	5.8
Yugoslavia	4.1	4.5	5.0	6.3	6.9	7.3
Cyprus/Gibraltar/Malta	—	—	—	—	—	—
Total Western Europe	**90.4**	**91.9**	**97.1**	**110.1**	**107.8**	**104.1**
Middle East	0.9	0.8	1.6	1.0	1.0	1.0
Africa	7.0	6.9	7.3	11.2	12.5	13.7
South Asia	8.1	8.0	7.4	10.2	10.7	12.8
South East Asia	6.3	6.4	7.3	7.7	8.0	8.6
Japan	21.9	17.3	19.1	17.4	19.2	20.7
Australasia	7.1	5.0	5.4	8.1	9.2	9.6
USSR	32.5	31.6	32.5	37.9	45.0	48.5
Eastern Europe	3.4	4.0	4.8	5.4	5.4	5.8
China	6.4	6.8	7.4	8.0	9.0	10.2
Total Eastern Hemisphere	**184.0**	**178.7**	**189.9**	**217.0**	**227.8**	**235.0**
World (excl. USSR E. Europe & China)	**275.3**	**286.1**	**309.3**	**319.8**	**348.6**	**352.3**
World	**317.6**	**328.5**	**354.0**	**371.1**	**408.0**	**416.8**

Table O. World Nuclear Energy Consumption 1971-1981 (MTOE)

Country/Area	*1971*	*1973*	*1975*	*1977*	*1979*	*1981*
USA	9.9	21.8	44.4	67.8	69.3	73.3
Canada	1.0	3.7	3.0	6.9	8.6	9.7
Total North America	**10.9**	**25.5**	**47.4**	**74.7**	**77.9**	**83.0**
Latin America	—	—	0.7	0.4	0.8	0.8
Total Western Hemisphere	**10.9**	**25.5**	**48.1**	**75.1**	**78.7**	**83.8**
Western Europe						
Austria	—	—	—	—	—	—
Belgium & Luxembourg	—	*a*	1.6	2.6	2.5	2.8
Denmark	—	—	—	—	—	—
Finland	—	—	—	0.6	1.6	3.5
France	2.3	3.0	3.9	5.1	8.3	22.1
Greece	—	—	—	—	—	—
Iceland	—	—	—	—	—	—
Republic of Ireland	—	—	—	—	—	—
Italy	0.9	0.8	0.9	0.9	1.2	1.3
Netherlands	0.1	0.3	0.8	1.0	0.8	1.0
Norway	—	—	—	—	—	—
Portugal	—	—	—	—	—	—
Spain	0.7	1.7	1.9	1.7	1.7	1.8
Sweden	*a*	0.5	3.0	4.9	3.6	6.3
Switzerland	0.7	1.6	1.9	2.1	2.4	3.7
Turkey	—	—	—	—	—	—
United Kingdom	7.1	5.9	6.3	8.4	8.1	8.1
West Germany	1.4	2.8	5.0	8.3	9.8	11.9
Yugoslavia	—	—	—	—	—	—
Cyprus/Gibraltar/Malta	—	—	—	—	—	—
Total Western Europe	**13.2**	**16.6**	**25.3**	**35.6**	**40.0**	**62.5**
Middle East	—	—	—	—	—	—
Africa	—	—	—	—	—	—
South Asia	0.5	1.0	1.0	0.8	0.6	0.9
South East Asia	—	—	—	—	1.6	3.5
Japan	2.1	2.3	5.3	6.9	14.9	21.2
Australasia	—	—	—	—	—	—
USSR	1.1	3.0	6.0	10.7	12.5	15.0
Eastern Europe	0.2	1.0	1.4	2.9	3.8	4.2
China	—	—	—	—	—	—
Total Eastern Hemisphere	**17.1**	**23.9**	**39.0**	**56.9**	**73.4**	**107.3**
World (excl. USSR E. Europe & China)	**26.7**	**45.4**	**79.7**	**118.4**	**135.6**	**171.9**
World	**28.0**	**49.4**	**87.1**	**132.0**	**152.1**	**191.1**

[a] Less than 0.05 million tonnes oil equivalent.

Table P. Approximate Conversion Factors

For Crude Oil[a]

FROM \ INTO	Tonnes	Long Tons	Short Tons	Barrels	Kilolitres (cub. metres)	1 000 Gallons (Imp.)	1 000 Gallons (US)
	MULTIPLY BY						
Tonnes (metric tons)	1	0.984	1.102	7.33	1.16	0.256	0.308
Long Tons	1.016	1	1.120	7.45	1.18	0.261	0.313
Short Tons	0.907	0.893	1	6.65	1.05	0.233	0.279
Barrels	0.136	0.134	0.150	1	0.159	0.035	0.042
Kilolitres (cub.metres)	0.863	0.849	0.951	6.29	1	0.220	0.264
1 000 Gallons (Imp.)	3.91	3.83	4.29	28.6	4.55	1	1.201
1 000 Gallons (US)	3.25	3.19	3.58	23.8	3.79	0.833	1

For Crude Oil and Products

TO CONVERT	Barrels to Tonnes	Tonnes to Barrels	Barrels/Day to Tonnes/Year	Tonnes/Year to Barrels/Day
	MULTIPLY BY			
Crude Oil[a]	0.136	7.33	49.8	0.0201
Motor Spirit	0.118	8.45	43.2	0.0232
Kerosine	0.128	7.80	46.8	0.0214
Gas/diesel	0.133	7.50	48.7	0.0205
Fuel Oil	0.149	6.70	54.5	0.0184

[a] Based on world average gravity (excluding natural gas liquids).

Table Q. Approximate Calorific Equivalents (oil = 10 000 kcal/kg)

One million tonnes of oil equals approximately:		*Heat units and other fuels expressed in terms of million tonnes of oil*	*Million tonnes of Oil*
Heat Units			
40	million million Btu	10 million million Btu approximates to	0.25
397	million therms	100 million therms approximates to	0.25
10 000	teracalories	10 000 teracalories approximates to	1.00
Solid Fuels[a]			
1.5	million tonnes of coal	1 million tonnes of coal approximates to	0.67
3.0	million tonnes of lignite	1 million tonnes of lignite approximates to	0.33
Natural Gas (1 cub. ft. = 1 000 Btu) (1 cub. metre = 9 000 kcal)			
1.111	thousand million cub. metres (BCM)	1 thousand million cub.metres (BCM) approximates to	0.90
39.2	thousand million cub. ft.	10 thousand million cub. ft. approximates to	0.26
0.0392	million million cub. ft. (TCF)	10 million million cub. ft. (TCF) approximates to	260
107	million cub. ft./day for a year	100 million cub.ft./day for a year approximates to	0.93
Electricity (1 kWh = 3 412 Btu) (1 kWh = 860 kcal)			
12	thousand million kWh	10 thousand million kWh approximates to	0.86

One million tonnes of oil produces about 4 000 million units (kWh) of electricity in a modern power station.

[a] Solid fuels: the equivalents now stated represent the relative calorific values of coal and lignite as produced.

Table R. OPEC Countries: Export Earnings (Millions of US $)

	1972	1973	1974	1975	1976
Algeria	1,306	1,896	4,683	4,691	5,201
Ecuador	343	548	1,135	987	1,293
Gabon	234	382	768	943	1,135
Indonesia	1,777	3,211	7,426	7,102	8,547
Iran	4,040	6,203	21,575	20,211	23,503
Iraq	1,100	2,193	6,956	8,301	7,854
Kuwait	2,557	3,321	10,325	8,644	9,844
Libya	2,477	3,454	7,129	6,042	8,306
Nigeria	2,178	3,525	9,699	7,776	10,085
Qatar	397	619	2,016	1,809	2,209
Saudi Arabia	4,560	7,747	31,242	27,996	36,437
UAE	1,082	1,807	6,392	6,970	8,684
Venezuela	3,126	4,891	11,071	8,880	9,299

	1977	1978	1979	1980	1981
Algeria	6,116	6,315	9,483	13,660	11,816
Ecuador	1,216	1,502	2,173	2,506	2,542
Gabon	1,343	1,107	1,848	2,173	n.a.
Indonesia	10,853	11,643	15,590	21,908	22,378*
Iran	24,260	22,200	19,876	14,278	9,443
Iraq	10,838	11,059	21,499	26,345	10,595
Kuwait	9,801	10,427	18,242	19,858	17,115
Libya	9,759	9,498	15,231	22,574	15,658
Nigeria	11,780	10,538	17,874	26,841	19,770
Qatar	1,992	2,374	3,779	5,672	n.a.
Saudi Arabia	44,061	37,914	58,751	102,113	113,328
UAE	9,637	9,126	13,652	20,747	20,037
Venezuela	9,545	9,188	14,317	19,221	20,950

*MID estimate.

Table S. OPEC Countries: Balance of Payments on Current Accounts (Millions of US $ (Minus sign: Deficits))

	1972	*1973*	*1974*	*1975*	*1976*
Algeria	-420	-817	198	-1,991	-888
Ecuador	-93	-22	9	-197	-6
Gabon	-12	-44	124	n.a.	-51
Indonesia	-421	-793	42	-1,136	-908
Iran	-392	86	10,916	n.a.	4,714
Iraq	543	810	2,855	2,970	2,627
Kuwait	†	†	†	†	6,950
Libya	392	266	2,037	-244	2,845
Nigeria	-320	45	4,995	474	-357
Qatar	n.a.	283	1,666	1,125	1,086
Saudi Arabia	1,757	2,903	23,634	18,085	13,799
UAE	n.a.	n.a.	n.a.	3,900	4,600
Venezuela	-56	696	5,794	2,745	256
OPEC	1,300	7,700	59,500	27,100	36,000

	1977	*1978*	*1979*	*1980*	*1981**
Algeria	-2,323	-3,538	-1,735	242	-2,000
Ecuador	-341	-701	-605	-747	-1,091
Gabon	3	74	245	600	0
Indonesia	-50	-1,414	980	2,850	-2,500
Iran	5,081	n.a.	n.a.	-1,000*	-4,000
Iraq	3,025	n.a.	n.a.	10,800*	-6,000
Kuwait	4,562	5,685	14,206	16,959	13,220
Libya	3,295	2,118	7,362	11,000	2,800
Nigeria	-1,019	-3,785	1,675	2,880	-7,000
Qatar	426	808	1,710	5,000	4,000
Saudi Arabia	11,905	-2,212	9,590	39,799	43,000
UAE	4,100	2,014	2,420	10,000	14,000
Venezuela	-3,079	-5,550	524	4,297	4,000
OPEC	29,000	4,000	62,000	105,000	60,000

*MID estimates.
†Figures not produced for individual years.

Table T. Current balances[a] **($ million)**

	1972	1973	1974	1975	1976
United States	-5,795	7,141	4,862[b]	18,280	4,384
Japan	6,624	-136	-4,693	-682	3,680
Germany	795	4,604	10,276	4,037	3,937
France	284	-967	-6,202	-240	-5,715
United Kingdom	617	-2,403	-7,656	-3,367	-1,583
Italy	2,043	-2,662	-8,017	-751	-2,816
Canada	-389	108	-1,493	-4,677	-3,897
TOTAL OECD	8,302	10,139	-25,061	2,387	-17,271

	1977	1978	1979	Average 1964–73	Average 1974–79
United States	-14,110	-14,075	1,414	211	126
Japan	10,918	16,534	-8,754	1,004	2,834
Germany	4,090	9,170	-5,262	1,317	4,375
France	-3,100	3,291	1,249	-288	-1,786
United Kingdom	-72	1,800	-1,828	180	-2,118
Italy	2,465	6,198	5,549	1,393	438
Canada	-4,043	-4,327	-4,179	-272	-3,769
TOTAL OECD	-24,780	9,808	-30,657	5,745	-14,262

[a] Goods, services and all transfer payments.
[b] Excluding cancellation of Indian debt (–1,993) and extraordinary grants (–746).

SECTION 7

ANNEXES

ANNEX I

Conference Programme

OPENING REMARKS

Sir William Hawthorne, Master of Churchill College

INTRODUCTORY ADDRESS

Sir Peter Baxendell, Chairman of Shell Transport and Trading

SESSION 1

World Oil and Gas Markets: The Changing Structure

Chairman: Alirio Parra, Director, Petroleos de Venezuela
Guest Speaker: Dr Ulf Lantzke, Executive Director, International Energy Agency
Speakers: M. A. Adelman, James Sweeney, Fereidun Fesheraki, Peter Odell
Rapporteur: B. Andrews, *Financial Times*

SESSION 2

The Response of Governments and the Business Sector

Chairman: Jane Carter, Under Secretary, UK Department of Energy
Guest Speaker: Michel Carpentier, Director General of the Commission of the European Communities
Speakers: John V. Mitchell, John H. Culhane
Rapporteur: H. Stonefrost, Bank of England

SESSION 3

The Energy Policy Perspective of the United States and the United Kingdom

Chairman: Paul Tempest, British Gas Corporation
Guest Speakers: The Hon. James B. Edwards, US Secretary for Energy
The Rt Hon. Nigel Lawson, UK Secretary of State for Energy
Rapporteur: R. C. Bending, University of Cambridge

First Dinner

Chairman: Mrs Jane Carter, Under-Secretary, UK Department of Energy
Speaker: The Rt Hon. Aubrey Jones, President of the Oxford Energy Policy Club

SESSION 4A

Coal, Nuclear and Gas

Chairman: Ian Smart, Ian Smart Ltd.
Speakers: Ravendra Pachauri, Fritz Lücke, H. Franssen, C. D. Kolstad
Rapporteur: Ian Smart

SESSION 4B

Energy Price Modelling

Chairman: Eric Price, Head of Economics and Statistics Division, UK Department of Energy
Speakers: Campbell Watkins, Colin Robinson, J. Wilkinson, H. D. Saunders, Len Brookes
Rapporteur: R. W. Byatt, National Westminster Bank

SESSION 4C

Oil and Gas Taxation, Tariffs and Licensing

Chairman:	R. O. Jackson, Group Oil Adviser, Midland Bank
Speakers:	Alexander Kemp, J. T. Fraser, Walter Mead, Laurence Jacobson
Rapporteur:	R. Dargie, Bank of England

SESSION 4D

Electricity, Alternative Energy, Technical Papers

Chairman:	Keith Williams, Shell International
Speakers:	S. A. Ravid, K. H. Rising, T. W. Berrie, N. Evans, I. C. Price
Rapporteur:	R. R. Jennison, British Nuclear Fuels

SESSION 5A

New Market Mechanisms in Oil and Gas

Chairman:	Dan Ion, Chairman of the World Petroleum Congress 1983
Speakers:	R. Netschert, Robert Deam, T. R. Dealtry, Joe Roeber, Niall Trimble
Rapporteur:	M. W. Clegg, BP

SESSION 5B

UK and US Department of Energy Presentations

Chairman:	John Barber, HM Treasury
Speakers:	K. Wigley, J. Stanley-Miller
Rapporteur:	R. C. Bending, University of Cambridge

SESSION 5C

The Economic Impact of Energy

Chairman:	Mariano Gurfinkel, Petroleos de Venezuela
Speakers:	Martha Krebs, Oystein Noreng, Elizabeth Parr-Johnston, Ira Sohn, T. Greening
Rapporteur:	H. Stonefrost, Bank of England

SESSION 5D

Australia and Canada

Chairman: Stuart Harris, Professor, RIIA and Australian National University, Camberra
Speakers: Russell S. Uhler, David Gallagher, G. D. McColl
Rapporteur: D. Russell, Consolidated Goldfields

SESSION 6

The Role of a State oil and Gas Company

Chairman: G. Corti, British National Oil Corporation
Guest Speaker: Jens Christensen, Chairman, Dansk Olie & Naturgas A.S.
Rapporteur: C. M. Buckley, BNOC

SESSION 7A

Global Strategic Issues

Chairman: Loren Cox, MIT
Speakers: A. F. A. Scanlan, John F. O'Leary, F. E. Banks, Robert Weiner
Rapporteur: L. Turner, RIIA

SESSION 7B

Energy Modellers Forum

Continuation of Sessions 4B and 5B

SESSION 7C

Energy Futures Markets

Chairman: Philip Warland, Bank of England
Speakers: John E. Treat, Shanta Devarajan, Glenn Hubbard, Walter Greaves, Thomas McKiernan
Rapporteur: A. Thorney, BP

SESSION 7D

OPEN — for any other specialist group wishing to discuss informally a topic not already on the programme.

Topic nominated:
Energy Economics in the Developing World

Speakers: Lee Schipper and others

SESSION 8

Responses to Disruption and Crisis

Chairman: Sir Archie Lamb, Samuel Montagu & Co.
Speakers: Robert Belgrave, Charles Ebinger, Philip Verleger
Discussant: Milton Russell, Resources for the Future
Rapporteur: R. M. Witcomb, BP

SESSION 9

The 1982 IAEE Prize Lecture

Chairman: William Hughes, Charles River Associates and President of the IAEE

Speaker: Sam Schurr, Electric Power Research Institute, Palo Alto

Topic: **Energy Efficiency and Productive Efficiency: Some Thoughts Based on American Experience**

SESSION 10

Demand Responses

Chairman: Campbell Watkins, President Datametrics Ltd
Speakers: Frank E. Hopkins, Lee Schipper, Leonard Waverman, Patrick O'Sullivan, Robin Wensley
Rapporteur: J. P. Prince, Royal Bank of Canada

Second Dinner

Chairman: Sir William Hawthorne, Master of Churchill College
Speaker: Michael Posner, Chairman of the Social Science Research Council

SESSION 11

Adaptation in the Developing World

Chairman: Helmut Frank, Professor, University of Arizona and

	Editor of the *Energy Journal*
Speakers:	James Plummer, C. M. Siddayao, Lutz Hoffman, Peter M. Meier, Marcus Fritz, C. Hope
Rapporteur:	L. Turner, RIIA

SESSION 12

Interfuel Substitution—Policy Opportunities in Changing Energy Markets

Chairman:	Joy Dunkerley, Resources for the Future and President-Elect of the IAEE
Guest Speakers:	Henry Jacoby, Melvin Conant, William R. Hughes, Robert Mabro, Jack Hartshorn, Christopher Johnson, Ray Dafter
Rapporteur:	Paul Tempest

ANNEX II

Delegate List

M. A. ADELMAN, Massachusetts Institute of Technology
B. ANDREWS, Financial Times (Press)
F. E. BANKS, University of Uppsala
J. M. BARBER, HM Treasury
Sir PETER BAXENDELL
C. H. BAYLY, Gaffney Cline and Associates
J. BEDORE, The Uranium Institute
R. BELGRAVE, British Institutes' Joint Energy Policy Programme
R. C. BENDING, University of Cambridge
Ms M. BERG, SMMT
E. R. BERNDT, Massachusetts Institute of Technology
T. W. BERRIE, Ewbank and Partners Limited
O. BGOERK, Swedish Embassy, London
S. BILLER, European Parliament
Miss E. J. BIRKBY, United Kingdom Atomic Energy Authority
C. C. BRADLEY, Department of Industry
L. G. BROOKES, United Kingdom Atomic Energy Authority
C. M. BUCKLEY, British National Oil Corporation
A. BUTCHER, Imperial Chemical Industries PLC
R. W. BYATT, National Westminster Bank Limited
Ms S. E. A. CANNEY, Burmah Oil Trading Limited
M. CARPENTIER, European Commission
Mrs J. CARTER, UK Department of Energy
B. du CHAFFAUT, Société Nationale Elf Aquitaine
J. H. CHESTERS, Watt Committee
J. CHRISTENSEN DONG
H. R. CLARKE, Asian Institute of Technology
M. W. CLEGG, BP Gas Limited
C. N. COLLYNS, Scallop Corporation
G. COMEL
M. A. CONANT, Conant and Associates Limited
Ms J. CONEY, Financial Times
F. P. COOKE, Esso Petroleum Company Limited
G. CORTI
L. C. COX, Massachusetts Institute of Technology
A. CREDE, European Banking Company
P. J. CRISSIUMA, Brazilian Embassy

J. H. CULHANE, Occidental Petroleum
R. DAFTER, Financial Times
G. R. DARGIE, Bank of England
J. R. DAVIES, Blue Circle Cement
G. R. DAVIS, Shell International Petroleum Company
S. DAY, Foreign and Commonwealth Office
T. R. DEALTRY, RBA Management Services Limited
R. DEAM, Queen Mary College, London
S. DEVARAJAN, Harvard University
G. DONNELLY, Midland Bank Limited
L. H. DRISCOLL, Butterworth Scientific
J. B. DUFFY, BP Oil Limited
Ms J. DUNKERLEY, Resources for the Future
R. E. EBEL, Greater Capital Area
C. K. EBINGER, Georgetown University
P. L. ECKBO, Petroplan International Incorporated
R. J. EDEN, University of Cambridge
J. EDWARDS, US Department of Energy
R. H. EDWARDS, Albright and Wilson Limited
R. T. EDWARDS, Bank of Scotland
G. H. G. ELLIS, Mobil Oil Company Limited
G. ESCOBEDO, Petroleos Mexicanos
J. K. EVANS, Pacific Resources Incorporated
N. EVANS, Cavendish Laboratory
N. R. FALLON
F. FESHARAKI, East-West Center
A. J. FINIZZA, Atlantic Richfield Company
P. C. FINSAAS
A. FLORES, Maraven S.A.
G. M. FOLIE, Shell Company of Australia Limited
P. A. FORD, Shell UK Limited
I. FORSYTH, Britannia Arrow Holdings PLC
K. N. FOSTER, BP Oil Limited
H. J. FRANK, University of Arizona
H. FRANSSEN, IEA
J. T. FRASER, Tennessee Gas Transmission
M. S. FRASER, Bank of Scotland
M. FRITZ, Max Planck Institute
M. FUCHS, Ministry of Finance
D. GALLAGHER, University of New South Wales
M. GARRICK, US Department of Energy
W. GREAVES
T. GREENING, First National Bank of Chicago
G. E. GRIFFIN, Mobil Oil Company Limited
M. GURFINKEL, Petroleos de Venezuela
Ms S. HALTMAIER TODD, Chase Manhattan Bank
A. HANSEN, Dansk Olie and Naturgas A.S.
S. HARRIS, The Royal Institute of International Affairs
J. HARTSHORN, Jensen Associates Incorporated
C. G. HARVEY, Shell International Petroleum Company Limited
M. E. HILDESLEY, Morgan Grenfell Energy Services
L. HOFFMANN, University of Regensburg
R. HOGAN, Construction News
A. B. HØNNINGSTAD, Royal Norwegian Embassy
C. HOPE, Cavendish Laboratory
F. HOPKINS, ORI Incorporated
Ms R. HOUGHTON, Financial Times
G. HUBBARD, Harvard University
W. R. HUGHES, Charles River Associates
N-O. IHLEN, Statoil
T. INOUE, The Tokyo Electric Power Company Incorporated
D. C. ION, World Petroleum Congress
R. A. IRVING, Petroleos de Venezuela
Ms W. JABLONSKI, Petroleum Intelligence Weekly
R. O. JACKSON, Midland Bank Limited
L. JACOBSON, Federal Reserve Board
H. D. JACOBY, Massachusetts Institute of Technology
L. JARASS, ATW GmbH
G. JARJOUR, U.P.M.
R. R. JENNISON, British Nuclear Fuels Limited
E. JOHNSEN, Norwegian Ministry of Petroleum and Energy
C. JOHNSON, Lloyds Bank Limited
The Rt Hon. A. JONES, Oxford Energy Club
C. L. JONES, UK Department of Energy
B. A. KALYMON, University of Toronto
J. KEEN-HOLLAND
A. KEMP, University of Aberdeen

S. KENNEDY, Texas Eastern International
H. KERRICH, College Retirement Equities Fund
P. Q. KING, United Kingdom Atomic Energy Authority
P. KNUTSSØN, Dansk Olie and Naturgas A.S.
C. D. KOLSTAD, Los Alamos National Laboratory
Ms M. KREBS, House Committee on Science and Technology
R. KRIJGSMAN, Petroleum Intelligence Weekly
Sir A. LAMB, Samuel Montagu and Company Limited
D. LANE, HM Diplomatic Service
U. LANTZKE, International Energy Authority
J. M. LAPORTE, Banque Nationale de Paris
The Rt Hon. NIGEL LAWSON
H. LOHNEISS
J. W. LONDON, BP Gas Limited
R. E. T. LONG
F. LUCKE, Ministry of Economics
R. MABRO, St Antony's College
G. D. McCOLL, University of New South Wales
J. McCORMICK, US Department of Energy
T. F. McKIERNAN, Drexel Burnham Lambert
L. P. MAGRILL, Texaco Limited
J. MALASHOCK, US Senate, Budget Committee
A. MANDEL, Technion Research and Development Foundation
Ms C. MARSHALL, Financial Times
R. J. MARTIN, Ultramar Company Limited
M. MATTHEWS, Petronal
W. J. MEAD, University of California
P. MEIER, Brookhaven National Laboratory
A. J. MEYRICK, UK Department of Energy
Ms S. MILLER, McGraw Hill World News
J. V. MITCHELL, British Petroleum Company Limited
A. MITTELSTADT, O.E.C.D.
M. MIYAIRI, Mitsui and Company
T. MONROE
J. MOORE, US Department of Energy
J. R. MORGAN, British National Oil Corporation
J. B. MORTENSEN, Copenhagen University
A. MOSEIDJORD, University of California
B. C. NETSCHERT, National Economic Research Associates Inc.
W. L. NEWTON, Petroleum Economics Limited
H. G. NICHOLLS, Tosco Corporation
B. P. NOLAN, The Irish Gas Board
O. NORENG, Norwegian School of Management
D. NUNN, Norsk Hydro A.S.
P. R. ODELL, Erasmus University
J. F. O'LEARY, O'Leary Associates
P. E. O'SULLIVAN, University of Wales
A. L. OTTEN, Cabot Corporation
R. PACHAURI, West Virginia University
A. A. PARRA, Petroleos de Venezuela
Ms E. PARR-JOHNSTON, Shell Canada Limited
C. PERALTA, Petroleos Mexicanos
L. PETTIGREW, The Royal Bank of Canada
J. PLUMMER, Q.E.D. Research Incorporated
P. POLLAK, The World Bank
A. H. PORTER, European Banking Company
M. POSNER, Social Science Research Council
E.H.M. PRICE, UK Department of Energy
G. PRICE, Shell International Petroleum Company Limited
I. C. PRICE, Sir William Halcrow and Partners
J. P. PRINCE, The Royal Bank of Canada
Colonel W. K. PRYKE
H. M. C. QUICK, Shell International Petroleum Company
D. C. RAO, The World Bank
S. A. RAVID, Haifa and Rutgers University
Ms B. W. RAWLS, Sun Company Incorporated
J. M. W. RHYS, Electricity Council
Ms K. H. RISING, Pacific Northwest Laboratory
C. ROBINSON, University of Surrey
M. S. ROBINSON, Shell International Petroleum Company
J. ROEBER, Joe Roeber Associates
T. D. ROSS, Shell International Petroleum Company Limited
Ms D. L. RUSSELL, Consolidated Gold Fields PLC
M. RUSSELL, Resources for the Future

Y. SARZIN, The World Bank
H. D. SAUNDERS, Tosco Corporation
T. SAWA, Kyoto University
A. J. A. SCANLAN, British Petroleum International
L. SCHIPPER, Lawrence Berkeley Laboratory
S. H. SCHURR, Electric Power Research Institute
M. SELIGMAN, European Parliament
C. M. SIDDAYAO, The East-West Center
I. SMART, Ian Smart Limited
M. H. SMITH, Sun Exploration and Production Company
I. SOHN, New York University
P. SORENSON, Florida State University
J. STANLEY-MILLER, US Department of Energy
Ms H. STONEFROST, Bank of England
J. L. SWEENEY, Stanford University
E. SYMONDS, Petroleum Economist
W. P. S. TAN, National Nuclear Corporation
M. TELSON, US House of Representatives, Budget Committee
P. TEMPEST, Bank of England/ British Gas Corporation
A. THORLEY, BP Oil Limited
L. THULIN, Den Norske Creditbank
J. M. TOMLINSON, Dow Chemical Company Limited
J. E. TREAT, New York Mercantile Exchange
N. TRIMBLE, British Gas Corporation
L. TURNER, Royal Institute on International Affairs
R. UHLER, University of British Columbia
Lord VAISEY, LASMO
P. K. VERLEGER Jr., Yale University
S. VINES, The Observer
J. N. VOLD, Shell International Petroleum Company
P. J. WARLAND, Bank of England
G. C. WATKINS, Datametrics Limited
L. WAVERMAN, Institute for Policy Analysis
R. WEINER, Harvard University
R. WENSLEY, London Business School
N. A. WHITE, Norman White Associates
K. J. WIGLEY, UK Department of Energy
Ms C. WILCOCKSON, Cavendish Laboratory
J. W. WILKINSON, Sun Company Inc.
K. R. WILLIAMS, Shell International Petroleum Company Limited
R. M. WITCOMB, BP International Limited
P. L. WOICKE, Morgan Bank
J. C. WOOD-COLLINS, Arthur D. Little
P. N. WOOLLACOTT, Atomic Energy Research Establishment
S. WRIGHT, Taylor Woodrow Construction Limited

IAEE and BIEE

THE INTERNATIONAL ASSOCIATION OF ENERGY ECONOMISTS

80 South Early Street, Alexandria, VA 22304 703/823 6966

Founded in the United States in 1977, the IAEE now has almost 2,000 full members. There are 14 US Chapters and affiliated national institutes, chapters and associations in 9 other countries. *The Energy Journal* is published quarterly by the IAEE and also distributed to all members.

Council: William R. Hughes *(President)*, Joy Dunkerley, M. A. Adelman, James L. Plummer, Joen Greenwood *(Secretary)*, Brian Sullivan, Jack Wilkinson, Jane Carter, Linda Ludwig, Paul Tempest, Mariano Gurfinkel, Linda S. Rathburn, Michael L. Telson, Samuel Van Vactor, Campbell Watkins.

THE BRITISH INSTITUTE OF ENERGY ECONOMICS

9 St. James's Square, London SW1

Founded in 1978 in London as the UK Chapter of the IAEE, the BIEE aims to further the understanding of all aspects of energy economics on both a national and international level. It maintains an archive in Chatham House and runs a regular meetings, conference and research programme. Membership is fairly evenly divided between government departments, the nationalised industries, the private sector, the universities and the City.

Committee: Paul Tempest *(Chairman)*, Jane Carter *(Vice-President)*, John Barber *(Treasurer)*, Eric Price *(Deputy Chairman)*, Ian Smart *(Secretary)*, Gerry Corti, Robert Deam, Walter Greaves, Christopher Johnson, Tony Scanlan, Niall Trimble, Norman White, Philip Warland.

From 1983, Eric Price, Head of Economics and Statistics Division, UK Department of Energy, will take over the Chairmanship and Philip Warland, Bank of England, the post of Treasurer. The following appointments have also been made: John Barber *(Member for Research)*; Gerry Corti *(Member for Planning)*; Tony Scanlan *(Member for Meetings Programme)*; Paul Tempest *(Chairman of the 1984 Cambridge Conference).*

1982 CAMBRIDGE ENERGY CONFERENCE

IAEE Programme Committee

Joy Dunkerley	United States
Jane Carter	United Kingdom
Lutz Hoffman	Germany
Mariano Gurfinkel	Venezuela
Campbell Watkins	Canada

BIEE Organising Committee

Jane Carter	UK Department of Energy
Tony Scanlan	British Petroleum
Eric Price	UK Department of Energy
Philip Warland	Bank of England
Cynthia Wilcockson	University of Cambridge

Chairman (IAEE and BIEE Committees):

Paul Tempest	Bank of England/British Gas Corporation

Instructions for Speakers, Rapporteurs and Participants

(as published in the Conference Programme)

Speakers and Chairmen

Apart from the guest speakers, all speakers are asked to present the summary and conclusion of their papers in 10 minutes, with an absolute deadline of 15 minutes. The Session Chairmen are asked to start promptly, enforce the 10-15 minute principle, so that there is ample time for questions and discussion, and on no account to overrun the allotted time for the session. If, by any chance, any speaker is squeezed out, we will fit him into a subsequent session.

Cynthia Wilcockson will be responsible for speakers throughout the Conference. She will be on-hand in or near the Conference Office or the Seminar Room, which is adjacent to it and which will be used as a speakers' lounge. Please let her have on your arrival two copies of your paper for the Proceedings Volume/ Energy Journal and for the Session Chairmen. The deadline for amendments to papers for publication is 15th August 1982. We suggest that Chairmen and speakers assemble in the speakers' lounge 10-15 minutes before their session begins, to ensure that the overhead projectors, rooms etc. meet their requirements and for a last minute briefing. All Chairmen on Monday and Tuesday are asked to note down, in say 6-8 points, what they consider the most significant remarks in the presentations in their session and to let Cynthia Wilcockson of the Conference Office have them by 9.00 am on Wednesday. These reports will be used in the final discussion and for the proceedings.

Rapporteurs
To strengthen the final open discussion, and for inclusion and acknowledgement in the Proceedings Volume, we should like to have on record a note of the main points made in the debate or discussion in each session. We are therefore requesting all participants who are not already speaking or chairing sessions to select an appropriate session. Cynthia Wilcockson will ensure that we have one rapporteur for each session. Please hand her your notes (up to two sides, in manuscript if legible) before leaving (sessions 1-10 before 9.00 am on Wednesday).

Abstracts and Copies of Papers
Abstracts of the James Sweeney paper on World Oil and the Barbara Laflin Richard/Fereidun Fesheraki paper on Crude Oil Price Differentials are given in this programme. Abstracts which we have received covering other sessions are available on the noticeboards, at the relevant session and from the office.

If you require a copy of any particular paper immediately, please ask the speaker. Otherwise, please fill in the Copies of Paper form at the registration desk: a fee of £4.00 per paper will be levied in advance to cover duplication and despatch by first-class post. We regret that we do not have resources to arrange for copies to be provided in Churchill College. All Conference Papers will be registered and retained in the BIEE Archive in Chatham House, London, where copies (at £4.00 including postage) may be obtained unless the authors instruct us otherwise.

1984 CAMBRIDGE ENERGY CONFERENCE

The next IAEE/BIEE Cambridge Energy Conference will be again held in Churchill College, Cambridge on 9-11 April 1984. The theme will be *International Energy Policy—A European and Middle East Perspective.* The Chairman of Royal Dutch Shell, Mr L. C. Van Wachem, has agreed to open the conference and the same level of ministerial support and international agency participation has been promised. A feature of the conference will be specialist sessions organised by the German, Benelux, Scandinavian and OPEC affiliates of the International Association of Energy Economists.

In organisation and format, the 1984 Conference will follow closely the 1980 and 1982 Cambridge Conferences. As previously, there will be no need to advertise the conference and the delegate list will again be closed at the 250 mark.

Enquiries about registration and advance reservation of places (deposit £20) should be addressed to:

Mrs M. Harrison, European Study Conferences Ltd, Kirby House, 31 High Street, Uppingham, Rutland, Leics LE15 9PY, United Kingdom. Their telephone number is 057-282-2711 and telex number 341352. The VAT number is 121 8066 96.

Nominations of papers should be sent to the Chairman, 1984 Cambridge Energy Conference, BIEE, 9 St James's Place, London SW1. A Conference Proceedings volume, similar to this volume, will be published and also distributed to all delegates.

Index

A

B

H

I

V

W

Z